# GENETIC ANALYSIS OF TUMOUR SUPPRESSION

The Ciba Foundation is an international scientific and educational charity. It was established in 1947 by the Swiss chemical and pharmaceutical company of CIBA Limited—now CIBA-GEIGY Limited. The Foundation operates independently in London under English trust law.

The Ciba Foundation exists to promote international cooperation in biological, medical and chemical research. It organizes about eight international multidisciplinary symposia each year on topics that seem ready for discussion by a small group of research workers. The papers and discussions are published in the Ciba Foundation symposium series. The Foundation also holds many shorter meetings (not published), organized by the Foundation itself or by outside scientific organizations. The staff always welcome suggestions for future meetings.

The Foundation's house at 41 Portland Place, London, W1N 4BN, provides facilities for meetings of all kinds. Its Media Resource Service supplies information to journalists on all scientific and technological topics. The library, open seven days a week to any graduate in science or medicine, also provides information on scientific meetings throughout the world and answers general enquiries on biomedical and chemical subjects. Scientists from any part of the world may stay in the house during working visits to London.

Ciba Foundation Symposium 142



# GENETIC ANALYSIS OF TUMOUR SUPPRESSION

*A Wiley – Interscience Publication*

1989

JOHN WILEY & SONS

Chichester · New York · Brisbane · Toronto · Singapore

Published in 1989 by John Wiley & Sons Ltd, Baffins Lane,
Chichester, Sussex PO19 1UD, UK.

Suggested series entry for library catalogues:
Ciba Foundation Symposia

Ciba Foundation Symposium 142
x + 258 pages, 34 figures, 24 tables

***Library of Congress Cataloging in Publication Data***

Genetic analysis of tumour suppression.
p. cm. — (Ciba Foundation symposium ; 142)
Proceedings of a symposium held at the Ciba Foundation, London, July 19–21, 1988.
Editors, Greg Bock, Joan Marsh.
"A Wiley–Interscience publication."
Includes bibliographies and indexes.
ISBN 0 471 92299 4
1. Tumors—Genetic aspects—Congresses. 2. Gene expression—Congresses. 3. Oncogenes—Congresses. I. Bock, Gregory. II. Marsh, Joan. III. Ciba Foundation. IV. Tumour suppression. V. Series.
[DNLM: 1. Cell Transformation, Neoplastic—congresses. 2. Gene Expression Regulation—congresses. W3 C161F v. 142 / QZ 50 G3225 1988]
RC268.4.G435 1989
616.99′2042—dc19
DNLM/DLC
for Library of Congress 89–5372
CIP

***British Library Cataloguing in Publication Data***

Genetic analysis of tumour suppression.
1. Man. Cancer. Genetic factors
I. Bock, Greg II. Marsh, Joan III. Series
616.99′4042

ISBN 0 471 92299 4

Typeset by Inforum Typesetting, Portsmouth
Printed and bound in Great Britain
by The Bath Press, Bath, Avon

# Contents

*Symposium on Genetic analysis of tumour suppression, held at the Ciba Foundation, London 19–21 July 1988*

*Editors: Greg Bock (Organizer) and Joan Marsh*

# Participants

**J.M. Adams** Walter & Eliza Hall Institute of Medical Research, Post Office, Royal Melbourne Hospital, Victoria 3050, Australia

**A. Balmain** Wolfson Institute for Molecular Pathology, Beatson Institute for Cancer Research, Garscube Estate, Switchback Road, Bearsden, Glasgow G61 1BD

**W.F. Benedict** The Center for Biotechnology, Baylor Medical College, 4000 Research Forest Drive, The Woodlands, Texas 77381, USA

**Sir Walter Bodmer** Imperial Cancer Research Fund Laboratories, PO Box 123, Lincoln's Inn Fields, London WC2A 3PX

**W.K. Cavenee** Ludwig Institute for Cancer Research, Montreal Branch, 687 Pine Avenue West, Montreal, Quebec, Canada H3A 1A1

**C.M. Croce** The Wistar Institute, 36th Street at Spruce, Philadelphia 19104-4268, USA

**H. Harris** Sir William Dunn School of Pathology, University of Oxford, South Parks Road, Oxford OX1 3RE

**G. Klein** Department of Tumour Biology, Karolinska Institute, Box 60400, S-104 01 Stockholm, Sweden

**H.P. Klinger** Department of Genetics, Albert Einstein College of Medicine, 1300 Morris Park Avenue, Bronx, New York 10461, USA

**A.G. Knudson, Jr** Institute for Cancer Research, Fox Chase Cancer Center, Philadelphia 19111, USA

**C-C. Lau** (*Bursar*) Department of Pharmacology, The Chinese University of Hong Kong, Faculty of Medicine, Shatin, NT, Hong Kong

**N. Lydon** CIBA-GEIGY Ltd, CH-4002 Basle, Switzerland

**P.A. Marks** The Memorial Sloan-Kettering Cancer Center, 1275 York Avenue, New York 10021, USA

**C.J. Marshall** Section of Cell & Molecular Biology, Institute of Cancer Research, Chester Beatty Laboratories, Fulham Road, London SW3 6JB

**B. Mechler** FB 21/Biologie, Institut für Genetik, Johannes Gutenberg-Universität Mainz, Postfach 3980, Saarstrasse 21, D-6500 Mainz 1, Federal Republic of Germany

**M. Noble** Ludwig Institute for Cancer Research, (Middlesex Hospital/University College Branch), Courtauld Building, 91 Riding House Street, London W1P 8BT

**M. Noda** Laboratory of Molecular Oncology, Tsukuba Life Science Center, Institute of Physical & Chemical Research, Riken, 3-1-1 Koyadai, Tsukuba, Ibaraki 305, Japan

**Sir Gustav Nossal** The Walter & Eliza Hall Institute of Medical Research, Post Office, Royal Melbourne Hospital, Victoria 3050, Australia

**L. Sachs** Department of Genetics, The Weizmann Institute of Science, Rehovot 76100, Israel

**M. Schwab** Institut für Experimentelle Pathologie, Deutsches Krebsforschungszentrum, P 101949, Im Neuenheimer Feld 280, D-6900 Heidelberg 1, Federal Republic of Germany

**M.B. Sporn** Laboratory of Chemoprevention, Division of Cancer Etiology, National Cancer Institute, Building 41, Room C-629, National Institutes of Health, Bethesda, Maryland 20892, USA

**E.J. Stanbridge** (*Chairman*) Department of Microbiology & Molecular Genetics, California College of Medicine, University of California, Irvine, California 92717, USA

**D. Tarin** Nuffield Department of Pathology, University of Oxford, John Radcliffe Hospital, Level 4, Headington, Oxford OX3 9DU

**A.J. Van der Eb** Department of Medical Biochemistry, Sylvius Laboratories, University of Leiden, PO Box 9503, 2300 RA Leiden, The Netherlands

**H.E. Varmus** Department of Microbiology & Immunology, University of California, School of Medicine, San Francisco, California 94143, USA

**R.A. Weinberg** Department of Biology, Massachusetts Institute of Technology, Whitehead Institute for Biomedical Research, Nine Cambridge Center, Cambridge, Massachusetts 02142, USA

**J.A. Wyke** Wolfson Institute for Molecular Pathology, Beatson Institute for Cancer Research, Garscube Estate, Switchback Road, Bearsden, Glasgow G61 1BD

**H. Zur Hausen** Deutsches Krebsforschungszentrum, P 101949, Im Neuenheimer Feld 280, D-6900 Heidelberg 1, Federal Republic of Germany

# Introduction

E.J. Stanbridge

*Department of Microbiology and Molecular Genetics, California College of Medicine, University of California, Irvine, California 92717, USA*

The majority of the participants at this symposium have been engaged in some form of research dealing with the genetic analysis of cancer. The idea of a genetic basis for cancer goes back to the turn of the century, when Boveri contended that chromosomal changes play a major role in the cancer process. Support for this notion has waxed and waned over the years. The discovery of cellular oncogenes about 10 years ago was a tremendous boost to our understanding of the genetics of dominantly inherited cancer. It was from the pioneering studies of Henry Harris and George Klein, who are here at the symposium, that the phenomenon of tumour suppression was first established using the technique of somatic cell hybridization. Interest in that phenomenon of tumour suppression has also waxed and waned over the years for a variety of reasons, but it is now generally accepted. I think we are now at a stage where we can begin to talk with some coherence about the genetic basis of tumour suppression. In addition to evidence from somatic cell hybridization experiments, more recently evidence of a more molecular nature has been accumulating for the existence of tumour suppressor genes. This comes from studies such as restriction fragment length polymorphism analysis of human malignancies where there are clear-cut indications of loss of genetic information associated with certain human cancers. A key distinction between oncogenes and tumour suppressor genes is that the former are activated in some fashion (by mutation, rearrangement, amplification, etc.), whereas inactivation (by mutation, deletion, etc.) of the latter is necessary for malignancy to occur. One example of a tumour suppressor gene is the retinoblastoma gene, which has recently been cloned.

This meeting is not devoted exclusively to tumour suppression and tumour suppressor genes but also to oncogenes. We must keep in mind that one does not substitute for the other. Cancer is not a single disease, but a complex of diseases with multiple genes involved. There is clearly room for both oncogenes and tumour suppressor genes to play key roles in the induction of cancer and I am looking forward to active discussion concerning possible interactions between these diverse genetic elements.

In 1986 Michael Stoker chaired a meeting on growth factors at the Ciba Foundation. In his introduction he mentioned that he suffered from dysacronymia. He sent out a plea that people would use words to describe important entities such as growth factors, rather than acronyms. I sympathize with that point of view, but we must also bear in mind that words can also be dangerous when used as descriptors. Some people consider that the term oncogene is used rather promiscuously and, on occasion, inappropriately. Genes involved in tumour suppression have already attracted a plethora of descriptive terms, including tumour suppressor genes, anti-oncogenes, recessive oncogenes and emerogenes. One must hope that a consensus will be reached in the not too distant future for an appropriate term for these important genetic elements.

# Epidemiology of genetically determined cancer

Alfred G. Knudson Jr.

*Institute for Cancer Research, Fox Chase Cancer Center, Philadelphia PA 19111, USA*

*Abstract.* Dominantly heritable susceptibility is known for virtually every cancer. Susceptibility is typically restricted to one or a few tumours. For some tumours there appear to be at least two different predisposing conditions. Some mutant gene carriers survive to old age without developing the expected tumour(s). Some cases are new germline mutations. None of the conditions is very common, because of natural selection against gene carriers.

Two questions arise: What is inherited? What is the relationship between the hereditary and non-hereditary forms of the same tumour? Retinoblastoma is a prototypic tumour. Penetrance in humans is nearly complete by the age of five years in the heritable form, which usually affects both eyes. Rare cases in which there is a constitutional deletion of chromosomal band 13q14 permitted localization of the responsible gene. Tumour formation is clearly a rare event at the cellular level, suggesting the necessity of a second, somatic, event. The difference in ages at diagnosis between unilateral and bilateral cases also suggests that two somatic events occur in non-hereditary cases. One explanation is that the gene is recessive and the second event involves loss of the remaining normal allele by mutation, non-disjunction, deletion or somatic recombination. The normal allele may be regarded as anti-oncogenic.

*1989 Genetic analysis of tumour suppression. Wiley, Chichester (Ciba Foundation Symposium 142) p 3–19*

## General observations

That cancer can be inherited has been well known for more than a century. It now appears that virtually every kind of human cancer has been found in a dominantly inherited form. Although the collective hereditary cancers make up only a small fraction of the total cancer burden, they obviously pose important problems for affected families. They have also illuminated previously unknown mechanisms in carcinogenesis.

The study of hereditary cancer reveals some common features. One of these is that genetic penetrance is difficult to assess. For those cancers that characteristically affect adults there is a strong age dependence and death from competing causes often interferes with evaluation of penetrance. Under some circumstances, notably among children with heritable cancer, it is clear that penetrance can be incomplete. A second feature is that more than one kind of cancer may be observed in pedigrees, although predisposition is never to cancer generally. Usually one, or a small number of tumour types, is observed, sometimes in one person, sometimes in different family members. A predominant type may occur as a primary tumour more than once in one individual. A few syndromes predispose to several cancers. Neurofibromatosis of the von Recklinghausen type may be accompanied by one or more of a long list of tumours, including neurofibrosarcoma, fibrosarcoma, pheochromocytoma, meningioma and glioma. Another syndrome predisposes to breast cancer, soft tissue sarcomas, osteosarcoma, adrenocortical carcinoma, glioma and leukaemia.

A third feature is that more than one distinct hereditary form of a cancer may exist. One hereditary colon cancer is not associated with polyposis of the colon, distinguishing it from familial polyposis coli; affected families do not have both forms. One kind of hereditary breast cancer may also predispose to ovarian cancer, but it is obviously different from the syndrome of breast cancer associated with soft tissue sarcomas.

Most hereditary cancers produce at least some deaths before the end of the reproductive period and are therefore subject to natural selection. Recurrent germinal mutations maintain the responsible gene in a population in 'mutational equilibrium', according to the relationship, $q = \mu/s$, where $q$ is the gene frequency, $\mu$, the mutation rate, and $s$, the coefficient of selection ($s = 0$ for a normal wild-type individual; $s = 1$ for a completely lethal gene). Since mutation rates are generally of the order of $10^{-5}$ per locus per generation, or smaller, and since $s$ is usually equal to or greater than 0.1 for hereditary cancer, the gene frequencies are generally less than $10^{-4}$, or the frequency of heterozygotes is less than $2 \times 10^{-4}$. Only polyposis coli and neurofibromatosis are reported to have frequencies of this magnitude, although hereditary breast cancer (without sarcomas) and hereditary non-polyposis colon cancer may also occur as frequently. For uncommon cancers, such as retinoblastoma, the hereditary fraction may be high, since the non-hereditary cases are rare too. The hereditary cancers often occur at earlier ages than do the corresponding non-hereditary cancers. One result of this is that the hereditary fraction is often larger at an earlier age, and the carriers of the mutation may have relative risks thousands of times higher than those of the general population at those same ages. It is not surprising that the frequencies of the hereditary cancers are similar world-wide, since they are maintained everywhere by germinal mutation, even though the incidences of the non-hereditary forms of the same cancer may vary considerably.

There are no reports of the existence of presumed homozygotes for any of the genes responsible for hereditary cancers, so it remains unknown whether there would be more serious effects than are found in heterozygotes.

At least 50 different dominantly inherited cancers or cancer syndromes are known in humans. Afflicted families experience fear and uncertainty, and physicians and genetic counsellors, frustration. Scientists have been given a challenge and provided with an opportunity to discover the basis for the remarkable predispositions and to apply this knowledge to cancer generally.

## Prototypic hereditary cancers

### *Retinoblastoma*

One of the hereditary cancers, retinoblastoma, has been considered as a prototype (Knudson 1971, 1978) and has been particularly informative. This tumour affects approximately five per 100 000 children world-wide. Nearly all cases occur before the age of six years; it is sometimes found in new-born infants. In about 25–30 per cent of all cases both eyes are affected. Although only approximately five per cent of cases have a family history of this tumour, the fraction of all cases that is heritable is 35–40 per cent. This discrepancy is explained by the fact that all bilateral and 10–15 per cent of unilateral cases can produce affected children; those with a negative antecedent but positive subsequent family history evidently carry new germinal mutations. In most countries three per 100 000 children develop non-heritable retinoblastoma, and two per 100 000 develop the heritable form.

Occasional individuals have had an affected parent and an affected child but were themselves unaffected; they clearly carried the germinal mutation without developing a tumour. Such people are uncommon, so penetrance is very high, even if incomplete. This, and the fact that gene carriers can develop tumours either unilaterally or bilaterally, force the conclusion that the inherited mutation is not sufficient for oncogenesis. Carriers of the mutation can develop no tumour, one, two or even more. If each tumour results from a single somatic event that occurs in a stochastic fashion, then the number of tumours in different cases should follow a Poisson distribution. This was tested by counting tumours in eyes of bilateral cases and comparing

**TABLE 1 Numbers of tumours in 66 bilateral retinoblastoma cases (Knudson 1971)**

| *Number of tumours in one eye* | *Frequency expected for m = 3 (%)* | *Observed frequency (%)* |
|---|---|---|
| 1 | 43 | 53 |
| 2 | 33 | 26 |
| 3 | 17 | 14 |
| 4 | 6 | 6 |
| 5 | 1.8 | 1.5 |

the results with frequencies of different numbers expected for various Poisson means (Knudson 1971). Since the non-zero classes would add up to 95 per cent for $m = 3$, that mean was of particular interest, and, indeed, the frequencies calculated for $m = 3$ compared favourably with the observed frequencies for 66 cases (Table 1).

Considered at the cellular level, oncogenesis is an extremely rare outcome, since retinoblasts are precursors for approximately $10^8$ differentiated cells in the retina. Even if cells in the last division cycles are not capable of expressing oncogenicity, there are probably at least $10^6$ cells capable of doing so. If the mean number of events is three, then transformation must occur with a frequency of about $10^{-6}$ or less per cell division. This rate is similar to observed mutation rates for somatic cells *in vitro*, and is therefore compatible with the idea that the second event is a somatic mutation at a particular locus.

For hereditary cases it was concluded that two events are involved, one a germline mutation, one a somatic mutation. What then of the non-hereditary cases? If only three per 100 000 children develop this form, while the gene carriers develop a mean of three tumours, then the relative risk of one tumour that is imposed by the germline mutation is 100 000. So large a difference in risk could be explained by a requirement for two events in the non-hereditary cases as well, but with both being somatic. Such an idea was supported by an analysis of ages at diagnosis for the two forms (Knudson 1971). Bilateral cases, which are all heritable, occur at earlier ages than do unilateral cases, most of which are non-hereditary. Plots of cases not yet diagnosed ($y$) at various ages ($t$) showed that bilateral cases declined in accord with the equation $d\ln y/dt = -k$, consistent with their being due to a single event. On the other hand, the unilateral cases declined in a curvilinear fashion, consistent with two events.

Hethcote and I (1978) developed a mathematical model that related the incidence of a childhood cancer to cellular processes. The essential features were that the incidence of the non-hereditary form, $q(t)$, depended upon the number of mitoses in target stem cells, $a(t)$, by time $t$, and upon the mutation rate ($v$) converting these target stem cells to intermediate, once-hit cells, and the mutation rate ($\mu$) converting intermediate cells to twice-hit tumour cells. The target cells were visualized as descendants of a few committed precursor cells, whose number was designated $b(o)$. This number is of the order of 10, whereas the final number of mitoses, $a(\infty)$, is approximately $10^7$. The relationship,

$$q(t) = \mu v\{[a(t) + b(o)]\{\ln[a(t) + b(o)] - \ln b(o)\} - a(t)\},$$

was derived. The final incidence was approximated by the expression

$$q(\infty) = \mu v\, a(\infty)\{\ln[a(\infty)/b(o)] - 1\}$$

and was in accord with the observed incidence using values of the independent variables that fit the data for both hereditary and non-hereditary forms.

For hereditary cases, the mean number of tumours at age $t$, $m(t)$, is the product of $\mu$ and $a(t)$. From these equations the estimated fraction of cases not diagnosed by a given age was compared with observations and found to fit well for both hereditary and non-hereditary cases.

If two mutations are necessary, what might be their relationship, if any? A simple hypothesis is that they affect the two copies of some autosomal gene. The mutation would therefore be recessive in oncogenesis at the cellular level, although inheritance of predisposition by individuals would be dominant, the anomaly arising because of the different genotypes of host and tumour at the locus in question. Two predictions, concerning the first and second events, could be made from this hypothesis.

One prediction was that the first event should be the same for both hereditary and non-hereditary tumours. Testing this idea depended upon localizing the defect. This was possible because a few cases of retinoblastoma have constitutional deletions that include chromosomal band 13q14 (Knudson et al 1976, Yunis & Ramsay 1978). From studies of linkage between the genes for retinoblastoma and esterase D, it has been shown that hereditary cases without deletion also involve this band (Sparkes et al 1983). Furthermore, this site is at least partially deleted in some tumours of individuals with the non-hereditary form. Molecular analyses have shown that a substantial number of tumours without visible deletion have an abnormality at this site (Friend et al 1986, 1987, Fung et al 1987, Lee et al 1987). It seems reasonable to conclude that all cases of retinoblastoma involve mutation of 13q14.

The second prediction was that the same locus in the homologous chromosome is lost or mutated as a consequence of the second event. Four mechanisms were proposed for this event: local mutation, chromosomal deletion, non-disjunction and resultant chromosomal loss, or somatic recombination (Knudson 1978). One clinical observation favoured the idea that the second event involved the homologous gene rather than an unrelated gene. A comparison of deletion cases with non-deletion cases showed that the former are less frequently bilateral and seem to have a mean number of tumours approximately one-half that of non-deletion cases. This can be explained by the recessive hypothesis in the following way. If the second event in a deletion case involves loss of the whole normal homologous chromosome, or a large deletion, or somatic recombination, then the candidate tumour cell would be homozygously deleted for a chromosomal segment that might include one or more genes that are essential for the cell, i.e. recessively lethal. On the other hand, more second events would be tolerated by a cell in which the inherited change is not visible microscopically, thus increasing the probability of bilateral disease.

Two groups of investigators used esterase D to study this problem. Godbout et al (1983) found that some tumours expressed only one allele of the enzyme, while the host's non-tumorous cells had two active alleles, a finding that could be explained by chromosomal loss or deletion, or by recombination. A patient with a retinoblastoma studied by Benedict et al (1983) had a half-normal level of esterase D, presumably due to an occult deletion that included both that gene and the retinoblastoma gene. The tumour contained only one chromosome 13 and no esterase D, apparently due to loss of the normal chromosome 13. Cavenee et al (1983) used restriction fragment length polymorphisms (RFLPs) to study this problem and were able to show that any of the aforementioned mechanisms can operate in evoking tumour formation. Their finding of somatic recombination in one tumour was the first demonstration of that phenomenon in humans and showed the value of studying somatic genetic mechanisms in tumours, because tumours amplify rare events.

### *Other tumours*

Wilms' tumour, neuroblastoma and pheochromocytoma demonstrated some of the features of retinoblastoma (Knudson & Strong 1972a, b). Bilateral cases were again related to heritable forms, and the ages at diagnosis were earlier for hereditary and bilateral cases. Wilms' tumour was especially interesting because of a small fraction of cases with sporadic aniridia (Miller et al 1964). We proposed that this association could be caused by chromosomal deletions that embraced separate, but neighbouring, genes for aniridia and for Wilms' tumour. The subsequent discovery of such deletions permitted the assignment of a Wilms' tumour gene to chromosome band 11p13 (Francke et al 1979). The use of syntenic RFLPs permitted the demonstration that the kinds of second events were similar to those seen in retinoblastoma (Koufos et al 1984). Furthermore, the suppression of tumorigenicity of Wilms' tumour cells by an added chromosomal arm 11p is consistent with a critical role for recessive events in that segment in the origin of this tumour (Weissman et al 1987).

A problem arises in Wilm's tumour that is not apparent for retinoblastoma. This concerns two other congenital defects that predispose to Wilms' tumour, hemihypertrophy and the Beckwith-Wiedemann syndrome. In hemihypertrophy there is asymmetric growth in one or more parts of the body, including the face, the trunk and the extremities. One may suppose that it is caused by a somatic mutation in the developing embryo, with excessive growth as a consequence. The Beckwith-Wiedemann syndrome seems to be a related condition in that it is accompanied by large birth weight, excessive size of the tongue (macroglossia), and sometimes by hemihypertrophy. One possibility is that it is produced by either a germinal mutation or an embryonic mutation, perhaps at the same locus as that affected in simple hemihypertrophy. Both conditions predispose to Wilms' tumour and to two other tumours of children, hepatoblastoma and rhabdomyosarcoma (Miller et al 1964, Sotelo-

**TABLE 2 Putative chromosomal sites of anti-oncogenes**

| | |
|---|---|
| 1p | Neuroblastoma |
| 3p | Small cell carcinoma lung<br>Renal carcinoma |
| 5q | Colon carcinoma |
| 11p | Wilms' tumour<br>Bladder carcinoma |
| 13q | Retinoblastoma<br>Ductal carcinoma breast |
| 22q | Acoustic neuroma |

Avila et al 1980). On rare occasions Beckwith-Wiedemann syndrome has been found in more than one family member, whereas hemihypertrophy has not. However, Meadows et al (1974) reported an individual with the latter condition, without Wilms' tumour, who became the mother of three children with Wilms' tumour. It would seem that two different mutations predispose to Wilms' tumour, one at the locus affected by aniridia deletion that is not associated with growth disturbance, and one at the Beckwith-Wiedemann locus, which predisposes to rhabdomyosarcoma and hepatoblastoma as well. The mosaic nature of the effects of some of these latter mutations points to autonomous local growth control. This manifestation of the predisposition seems different from that expected of a recessive mutation, resembling more that expected of an oncogene mutation.

Osteosarcoma is another paediatric tumour that seems to involve a recessive gene. It occurs in about 10 per cent of the bilateral cases of retinoblastoma as a second cancer, and is clearly related to the heritable form of the eye tumour. A test of the hypothesis that the non-hereditary form of a tumour involves mutation at the same locus as in the hereditary form has been conducted with RFLPs on chromosome arm 13q, and has revealed loss of heterozygosity in a significant fraction of cases (Friend et al 1987).

A recessive mechanism has now been proposed for a number of cancers. Some of these and their presumed chromosomal locations are shown in Table 2. The list includes some common tumours of adults. Especially strong evidence has been found for colon cancer. As noted previously, two hereditary forms of this cancer are known, one with associated polyposis, one without. The polyposis gene has recently been mapped to chromosomal arm 5q (Bodmer et al 1987). This made possible a survey of non-hereditary colon cancers for loss of heterozygosity for syntenic RFLPs, which was found in about 25 per cent of cases (Solomon et al 1987). This is about what one would expect if non-hereditary cancers involved the polyposis locus in approximately one-half of tumours, and if the second events could be detected by such loss of heterozygosity in one-half of those tumours.

### Oncogenesis and development

Two distinct classes of human cancer genes have been identified. One class, the oncogenes, was discovered through the study of tumour viruses; the other, variously known as recessive oncogenes, tumour suppressor genes or anti-oncogenes, through the study of hereditary cancer. Both classes were anticipated by Theodor Boveri in 1914, in his famous treatise on the origin of cancer:

'. . . in every normal cell there is a specific *arrangement for inhibiting*, which allows the process of division to begin only when the inhibition has been overcome by a special stimulus. To assume the presence of definite *chromosomes which inhibit division*, would harmonize best with my fundamental idea. . . . Cells of tumours with unlimited growth would arise if those 'inhibiting chromosomes' were eliminated. . . . On the other hand, the assumption of the existence of chromosomes which *promote* division, might satisfy this postulate . . . cell-division would take place when the action of these chromatin parts . . . should be strengthened by a stimulus. . . . If three or four such chromosomes meet, the whole number of chromosomes being otherwise normal, then the tendency to rapid proliferation would arise.'

To date there have not been any demonstrations of heritable oncogene mutations that predispose to cancer, but the possibility that two oncogene abnormalities may be necessary for cancer induction suggests that such may exist. Perhaps the heterozygous state would manifest some disturbance in non-neoplastic growth, as is seen in the Beckwith-Wiedemann syndrome and in the phacomatoses, including von Recklinghausen's neurofibromatosis. The finding of a constitutional oncogene abnormality that strongly predisposes to cancer would greatly strengthen the evidence that such mutations play a role in the initiation of human cancer.

It appears likely that many cancers contain abnormalities in both oncogenes and anti-oncogenes. Certainly many human tumours show mutations in one of the *ras* oncogenes, and/or amplification of one member of the *myc* family of oncogenes. Much less is known about the suppressor genes, since only one of the human genes has been cloned. The high frequency with which tumours show loss of heterozygosity for various markers, and the well known observation that the oncogenicity of most human tumours is suppressed by hybridization with normal cells, support a high frequency of recessive genes in oncogenesis. However, it has not been shown for any tumours, other than some paediatric ones, that the remaining allele is actually mutated. This is where the hereditary cancers can be very useful. Similarly, suppression of oncogenicity in hybrids is not a guarantee of the primacy of a recessive mutation. Thus, it may be possible that growth of a tumour with a mutant oncogene could be suppressed by four copies of some regulatory gene, but not by the two copies normally present.

**TABLE 3 Hereditary conditions that predispose to oncogenesis in neural crest derivatives**

| |
|---|
| von Recklinghausen's neurofibromatosis |
| Central neurofibromatosis |
| Hereditary neuroblastoma |
| Hereditary pheochromocytoma |
| Hereditary medullary carcinoma of thyroid |
| Multiple endocrine neoplasia type 2 |
| Hereditary chemodectoma |
| Dysplastic nevus syndrome |

The number of tumour suppressor genes may be large. There are at least 50 distinguishable hereditary cancers, each possibly attributable to a different gene. What is the normal function of this class of gene? The specificities of tumour formation, even when more than one tumour results, suggest a role in differentiation. So too do the examples of more than one hereditary cancer in developmentally related tissues. The most outstanding example of this concerns disorders of the tissues derived from neural crest where as many as eight different conditions can effect tumorigenesis in one or more of these tissues (Table 3). In the case of the adrenal medulla gland there are three different conditions that produce tumours: hereditary neuroblastoma, hereditary pheochromocytoma and multiple endocrine neoplasia type 2. Here the normal allele of the responsible gene seems to be specific not only for the tissue, but also for the state of differentiation within the tissue.

*Acknowledgements*

Supported by grants CA-06927 and CA-43211 from the United States Public Health Service and by an appropriation from the Commonwealth of Pennsylvania.

## References

Benedict WF, Murphree AL, Banerjee A, Spina CA, Sparkes MC, Sparkes RS 1983 Patient with 13 chromosome deletion: evidence that the retinoblastoma gene is a recessive cancer gene. Science (Wash DC) 219:973–975

Bodmer WF, Bailey CJ, Bodmer J et al 1987 Localization of the gene for familial adenomatous polyposis on chromosome 5. Nature (Lond) 328:614–616

Boveri T 1914 Zur Frage der Entstehung maligner Tumoren. Gustav Fischer, Jena; English translation, The origin of malignant tumours, by M. Boveri, Williams and Wilkins, Baltimore, 1929

Cavenee WK, Dryja TP, Phillips RA et al 1983 Expression of recessive alleles by chromosomal mechanisms in retinoblastoma. Nature (Lond) 305:779–784

Francke U, Holmes LB, Atkins L, Riccardi VM 1979 Aniridia-Wilms' tumor association: evidence for specific deletion of 11p13. Cytogenet Cell Genet 24:185–192

Friend SH, Bernards R, Rogelj S et al 1986 A human DNA segment with properties of the gene that predisposes to retinoblastoma and osteosarcoma. Nature (Lond) 323:643–646

Friend SH, Horowitz JM, Gerber MR et al 1987 Deletions of a DNA sequence in retinoblastomas and mesenchymal tumors: organization of the sequence and its encoded protein. Proc Natl Acad Sci USA 84:9059–9063

Fung Y-KT, Murphree AL, T'Ang A, Qian J, Hinrichs SH, Benedict WF 1987 Structural evidence for the authenticity of the human retinoblastoma gene. Science (Wash DC) 236:1657–1661

Godbout R, Dryja TP, Squire J, Gallie BL, Phillips RA 1983 Somatic inactivation of genes on chromosome 13 is a common event in retinoblastoma. Nature (Lond) 304:451–453

Hethcote HW, Knudson AG 1978 Model for the incidence of embryonal cancers: application to retinoblastoma. Proc Natl Acad Sci USA 75:2453–2457

Knudson AG 1971 Mutation and cancer: statistical study of retinoblastoma. Proc Natl Acad Sci USA 68:820–823

Knudson AG 1978 Retinoblastoma: a prototypic hereditary neoplasm. Semin Oncol 5:57–60

Knudson AG, Strong LC 1972a Mutation and cancer: neuroblastoma and pheochromocytoma. Am J Hum Genet 24:514–532

Knudson AG, Strong LC 1972b Mutation and cancer: a model for Wilms' tumor of the kidney. J Natl Cancer Inst 48:313–324

Knudson AG, Meadows AT, Nichols WW, Hill R 1976 Chromosomal deletion and retinoblastoma. N Engl J Med 295:1120–1123

Koufos A, Hansen MF, Lampkin DB et al 1984 Loss of alleles at loci on human chromosome 11 during genesis of Wilms' tumour. Nature (Lond) 309:170–172

Lee W-H, Bookstein R, Hong F, Young L-J, Shew J-Y, Lee EY-HP 1987 Human retinoblastoma susceptibility gene: cloning, identification, and sequence. Science (Wash DC) 235:1394–1399

Meadows AT, Lichtenfeld JL, Koop CE 1974 Wilms' tumor in three children of a woman with congenital hemihypertrophy. N Engl J Med 291:23–24

Miller RW, Fraumeni JF, Manning MD 1964 Association of Wilms' tumor with aniridia, hemihypertrophy and other congenital malformations. N Engl J Med 270:922–927

Solomon E, Voss R, Hall V et al 1987 Chromosome 5 allele loss in human colorectal carcinomas. Nature (Lond) 328:616–619

Sotelo-Avila C, Gonzales-Crussi F, Fowler JW 1980 Complete and incomplete forms of Beckwith-Wiedemann syndrome: their oncogenic potential. J Pediatr 96:47–50

Sparkes RS, Murphree AL, Lingua RW et al 1983 Gene for hereditary retinoblastoma assigned to human chromosome 13 by linkage to esterase D. Science (Wash DC) 219:971–973

Weissman BE, Saxon PJ, Pasquale SR, Jones GR, Geiser AG, Stanbridge EJ 1987 Introduction of a normal human chromosome 11 into a Wilms' tumor cell line controls its tumorigenic expression. Science (Wash DC) 236:175–180

Yunis JJ, Ramsay NKC 1978 Retinoblastoma and subband deletion of chromosome 13. Am J Dis Child 132:161–163

## DISCUSSION

*Klinger:* Al (Knudson), your mathematical models require at least two events—can they predict a maximum number of events?

*Knudson:* No, because a three event model can be made to fit virtually any curve.

*Noble:* If you do transfection experiments with genomic DNA or cDNA from retinoblastoma cell lines, can you pick up *ras* type or *myc* family type oncogenes?

*Weinberg:* We have not succeeded in that, nor do I know of anyone else who has. In spite of this, I am reluctant to believe that inactivating both copies of the *Rb* gene is sufficient to generate a full blown retinoblastoma. I discussed this issue with Al Knudson 4–5 years ago, expressing the belief that something else is required for a full blown tumour. I suspect that inactivation of both *Rb* alleles may yield a benign localized lesion, but something else is needed to push cells further into the fully malignant state. I know Al feels that inactivation of the two copies is sufficient for a tumour.

*Knudson:* Allow me to make a qualification. A mathematical model can deal only with rate-limiting events; it has nothing to do with the absolute number of events that one can find in a tumour. For example, suppose two events permit growth to a million cells and then the tumour stalls: some further event is necessary. If that event occurs at a frequency of $10^{-4}$/cell division, it would count as a two event tumour because it is inevitable that the event would occur. So the biologist would think it is important, but in the mathematical model it doesn't count at all.

*Harris:* I suppose one would define a secondary event as one that may or may not occur but which is not necessary for the development of the tumour. If these other things we are talking about are secondary events, then one would expect to see retinoblastomas that are essentially euploid except for the two lesions conferring homozygosity at the one locus. When I ask people what the actual experience is, I get totally conflicting answers. Are retinoblastomas commonly euploid apart from the homozygous recessive?

*Knudson:* I think they are almost always aneuploid.

*Harris:* In that case, there must be at least three events.

*Knudson:* Again I come back to the question of whether they are rate limiting.

*Harris:* I agree with your analysis, but if, in addition to the homozygous recessives, some other scrambling event that grossly alters the karyotype invariably occurs, then there is a problem of semantics.

*Knudson:* I warned that the deeply buried events could be important to get the tumour started or they could be part of the progression from a benign form to a malignant form—so it is a semantic problem.

*Harris:* I recall a paper by Bonaïti-Pellie et al (1976) in which the authors present another form of mathematics that fits three events better than two. I assume that again raises the question of what you call a second event.

*Knudson:* The model that Hethcote and I made uses their data.

*Harris:* So you can fit two events to the Bonaïti-Pellie data?

*Knudson:* Yes.

*Adams:* Are there any developmental abnormalities associated with retinoblastoma, other than the tumours themselves?

*Knudson:* No, the patients with a deletion can have other abnormalities but that's presumably due to deletion of other genes. The mutations of the *Rb* gene are not associated with developmental abnormality and I would say that is, without exception, true for the hereditary cancers. The only time one sees abnormalities is with deletion. As you know, the 5q location of the polyposis gene was suspected because of anomalies associated with a case of polyposis in which there was a deletion in 5q.

*Stanbridge:* You mentioned that the loss of one allele may lead to a cell altered in some way, although not a tumour cell, and you gave the example of polyposis. Are there any pathological symptoms in the retinas of bilateral cases?

*Knudson:* Some people say that they can see little nests of retinoblasts that they don't think are tumours, especially peripherally in the retina. These are seen in the bilateral cases but not in the unilateral cases. I don't think anybody has ever looked at this systematically.

*Benedict:* Brenda Gallie has looked at the benign form of the disease. Since these tumours have not really been isolated and put back in culture, it is not possible to know what these lesions represent. You described how another event may take a malignant cell from its tumorigenic state to a completely differentiated form. This may be what has happened in these cells. We have tried to grow such cells, but have not yet succeeded. Whereas retinoblastoma is very malignant once it gets outside the eye, these benign cells only form small nodules and may be like nephroblastoma, which can completely differentiate *in vivo*.

*Knudson:* Yes, Brenda Gallie refers to this as retinoma, a benign lesion. She favours the idea that the second event may even be the same as the one found in retinoblastoma but it has occurred very late in the development of the retina. Even though the cell has acquired the second critical event, it has limited mitotic capacity and does not become malignant.

Intermediate lesions of the type seen in polyposis are known in some other heritable cancers. For example, in medullary carcinoma of the thyroid one can see hyperplasia all through the medulla in the hereditary form but not in the non-hereditary form. Is this hyperplasia the result of the first, hereditary event or has it involved a second event? In polyposis, is the polyp an offshoot or is it on the path to carcinoma? There is an argument among pathologists: some pathologists insist that one can find carcinomas arising from the flat epithelium in polyposis and that the polyp is not a necessary intermediate.

*Klinger:* I have also heard from pathologists that they have the impression that in the non-polyp epithelium there are cancerous changes. There is always the problem with histological sections of knowing just what part of the tissue you are looking at.

*Knudson:* A scenario that would be interesting to examine is the following: the original epithelium undergoes some *ras*-like mutation, so that there is much more growth, but not malignancy. It would be more likely that a carcinoma would arise from such a cell just because there are more of them, but the *ras* mutation would not be a necessary part of the generation of that carcinoma.

*Schwab:* Evaluating the role that the deletion might play in the development of the malignant tumour, is anything known about secondary changes, non-random chromosomal changes in the retinoblastoma cells? In some cases there is gene amplification and there could be other secondary changes involved in the generation of the malignant phenotype.

*Knudson:* Dr Kuznetsova in the Soviet Union and Brenda Gallie found an iso6p chromosomal alteration in 60% of the cases of retinoblastoma. I think they tend to regard it as something that makes a better retinoblastoma but isn't necessary.

*Benedict:* The most common chromosomal change seen in retinoblastoma is the iso6p and it has not been seen in many other tumours. It is found in a large percentage of retinoblastomas. There is also an increased representation of the long arm of chromosome 1, but this is found in many kinds of tumours. It is thought to be a secondary change, possibly connected with an increased capacity for proliferation. Initially, we found that the N-*myc* gene was highly expressed in all the retinoblastomas that we looked at. We were quite excited by this but when we went back and looked at normal fetal retinal cells, we found that the N-*myc* gene was highly expressed there as well. It may be that the N-*myc* gene is important in the early development and proliferation of the retina and is a marker of the developing retinoblast.

*Schwab:* Is there also some genetic change or gene amplification in those cases?

*Benedict:* In most cases, the increased expression of N-*myc* found in retinoblastoma was not associated with gene amplification. N-*myc* amplification, as you know, is seen primarily in neuroblastoma.

*Klein:* Dr Knudson, would you speculate on the role of these genes in differentiation. In your early papers, when there was essentially only the retinoblastoma story, you suggested that the *Rb* gene may be needed for a terminal differentiation step of the retinoblast. That became more complicated when the association of the *Rb* gene with osteosarcoma was discovered. Clearly, the second allele must be lost frequently in all dividing tissues, not only in the progenitors of retinoblastoma and osteosarcoma. Why do tumours not arise in many other tissues? Isn't there a similar situation in the Wilms' tumour, where tumours arise only in certain specific tissues? Can you think of pleiotropic controls that control differentiation in some tissues but not in others?

*Knudson:* One possibility is that the critical second event occurs only in certain tissues. Another possibility is that the event occurs in many tissues, including some that never express a tumour, where it is not critical, perhaps because other growth controls override the deficiency.

Loss of heterozygosity for 13q markers has been observed in some cases of breast cancer, yet breast cancer is not observed in retinoblastoma carriers more frequently than in the general population. Perhaps the critical site is not the *Rb* gene or additional changes may be necessary for breast cancer. It is unlikely that the *Rb* gene is the site of mutation in hereditary breast cancer, because penetrance is very high in the latter.

*Weinberg:* We are confronted with at least three paradoxes at present. One stems from the fact that the *Rb* gene is inactivated somatically in several kinds of tumours, including small cell carcinoma of the lung and mammary carcinomas, that do not appear as secondary tumours in children who have a retinoblastoma early in life. The second is that the *Rb* gene is expressed in many cell types, yet only a small number of cell types yield tumours upon homozygous inactivation of their *Rb* genes. The third paradox is that so far the disease of retinoblastoma has only been documented in human beings, even though we believe that the genes controlling development and tissue formation are common to all mammals. Thus there are three issues of tissue specificity and species specificity for which we have no answers at present.

*Benedict:* In collaboration with Drs Stanbridge and Weissman, we have done some experiments introducing a normal chromosome 13 into tumour cell types in which we believe that loss of the *Rb* gene is involved. In our initial study with the osteosarcoma, OSV, this has resulted in tumour suppression, which suggests that we are going in the right direction.

There is also information on changes in the *Rb* gene in several other human tumours. Dr Cavenee and his collaborators found that in hereditary breast cancers there was a change from heterozygosity to homozygosity for chromosome 13 in the cancer cells. Both Jun Yokota and Sue Naylor have found similar changes in small cell carcinoma of the lung, although 3p was also deleted in these tumours. Subsequently, deletions in the *Rb* gene have been found in several breast cancers. The first to find such changes was Dr Fung with his colleagues (T'ang et al 1988). In addition, both Dr Minna's group and Dr Yokota in collaboration with us have found abnormalities in the *Rb* gene in several cases of small cell carcinoma of the lung (Harbour et al 1988, Yokota et al 1988).

The published findings of Dr Weinberg and his colleagues on deletions within the *Rb* gene in the soft tissue sarcomas don't prove that the *Rb* gene is necessarily involved in these tumours, although this is likely to be the case. All the cases reported were of total loss of the gene or a deletion from the 5′ end. Therefore there could be a gene 5′ to the lesion that is involved in these particular tumour types. However, in the case of breast cancer and small cell carcinoma of the lung, as well as retinoblastoma and osteosarcoma, *internal* deletions have been found. I think in these cases the *Rb* gene plays a role in tumorigenicity but it is just one of the genes that is important.

Dr Stanbridge and his colleagues have developed a mouse–human cell line that contains a human chromosome 13 carrying a neomycin resistance marker gene. This can now be used to investigate whether, when it is put back into small cell carcinoma of the lung, breast cancer, retinoblastoma or another of these tumour types, they revert to non-tumorigenic cells. Such a result would imply that the *Rb* gene itself is involved in the formation of these various tumour types. The difference would be that in retinoblastoma possibly only two lesions are necessary, the loss of both *Rb* alleles, whereas loss of function of the *Rb* gene is only one of several functions required for the development of breast cancer. The role of the *Rb* gene in these tumours may be very significant, however.

*Weinberg:* With respect to the work on soft tissue sarcomas, Steve Friend has looked at a number of different sarcomas, including histiocytoma and several soft tissue sarcomas. The number of gross deletions that occurred homozygously in the *Rb* gene was three out of 13 cases. That's about the rate at which one observes deletions of the *Rb* gene in retinoblastomas by Southern blotting. Therefore it is conceivable, although hardly proven, that in virtually all of these soft tissue sarcomas there was a lesion in the *Rb* gene. Ultimately, the only way to verify this is to check for the presence or absence of functional protein.

*Stanbridge:* Did those lesions extend 5′ of the gene?

*Weinberg:* I think in one histiocytoma, which is supposedly a sarcoma, there was a fully internal deletion, but in general one could not tell whether the lesion was in the *Rb* gene or in an adjacent gene.

I only want to leave the speculation that in all sarcomas which arise as a consequence of somatic mutation (as opposed to some germline predisposition), there may be homozygous inactivation of the *Rb* gene. It may be as frequent as in retinoblastomas.

*zur Hausen:* Bob mentioned the interesting point concerning the absence of retinoblastomas in animals. Is that accepted? Are there any data on *in vitro* transformation of retinoblasts from animals?

*Weinberg:* I don't know about *in vitro* transformation of the retinoblasts of many types of animals; I think that it has been done in the eye of the chicken. However, spontaneously arising retinoblastomas in domestic animals or in mice or rats have never been documented. That suggests there may be something subtly different between the genes that regulate photoreceptor cone development in humans and in other species.

*Benedict:* I believe there is retinoblastoma in the wild possum. Unfortunately, you can't breed them in captivity, so we could not do genetic studies.

*Knudson:* There are induced retinoblastomas, they can be induced by adenovirus.

*Weinberg:* It's not clear that those are exact phenocopies of the spontaneously arising tumours. They may not even come from the same lineage of retinal cells.

*Klinger:* The various points that have been made about the *Rb* gene suggest that it may belong to the class of genes which exert their main effect during a specific time-frame of differentiation. Such genes, including oncogenes, have been shown to act in this way in studies of gene expression during normal development, as well as in studies with transgenic mice. If this is true, then introducing a chromosome 13 from a normal human cell, or the cloned *Rb* gene, into retinoblastoma cells might not suppress their tumorigenicity, even though the correct 'suppressor' is there.

*Marks:* In patients homozygous for the deletion of the *Rb* gene, is there any evidence for differentiation of the cells in their retinoblastoma tumours or are they homogenous with regard to cell type?

*Benedict:* They cover a wide range. There is a histological pattern known as Flexner-Wintersteiner rosettes, which are used as the basis for the pathological diagnosis of a differentiated retinoblastoma. Within a tumour you can find areas that are highly differentiated, with these rosettes, and areas that are not: both have the same chromosome composition.

*Klinger:* Without any evidence for deletion of the *Rb* gene?

*Benedict:* There is evidence that it has been deleted in both cases. You can put them in culture and with the correct conditions they will give rise to cultures containing almost 100% Flexner-Wintersteiner rosettes.

*Marks:* It appears that under those circumstances, deletion of the *Rb* gene is not sufficient to prevent differentiation of the cells.

*Benedict:* They are not terminally differentiated, they still divide. What I find interesting is that clinically retinoblastoma, if it gets outside the eye, never gives rise to Flexner-Wintersteiner rosettes. The same thing happens in the nude mouse: if you put differentiated tumour cells into the eye of a nude mouse, there is differentiation; if you inject them subcutaneously, you lose differentiation. If you then remove the subcutaneous tumours and put them back into the eye, you never get a differentiated cell again. So once the cell leaves the eye, it receives a signal that tells it never to differentiate again.

*Cavenee:* That is probably just selection for less differentiated subclones.

*Benedict:* No, because you can do it in culture as well. They are not terminally differentiated, these cells continue to proliferate.

*Cavenee:* As rosettes?

*Benedict:* Yes, as rosettes, both in the eye and in culture. Cytogenetically, they have a similar pattern and are very stable karyotypically, containing the same marker chromosomes. But once you take them out of the eye they never give rise to differentiated tumours again, nor will they form rosettes in culture, unlike neuroblastoma.

*Sachs:* How do you know you are not selecting non-differentiable cells? You have a mixed population to begin with.

*Benedict:* You can get differentiation in the eye after injecting 6000 or fewer cells. In the eye you get the same tumour pattern that was originally seen in the primary tumour, i.e. both differentiated cells and non-differentiated cells.

These tumours arise very rapidly, and it is hard to believe that selection could operate that fast. Karyotypically and genetically they look the same: they have lost the *Rb* gene and if there is a new marker chromosome, they have the same marker chromosome in the differentiated and non-differentiated cell. The issue is what the Flexner-Wintersteiner rosette really represents—it is a feature of differentiation but certainly not of terminal differentiation.

## References

Bonaïti-Pellie C, Briard-Guillemot ML, Feingold J, Frezal J 1976 Mutation theory of carcinogenesis in retinoblastoma. J Natl Can Inst 57:269–276

Harbour JW, Shinn-Liang L, Whang-Peng J, Gazdar AF, Minna JD, Kaye FJ 1988 Abnormalities in structure and expression of the human retinoblastoma gene in SCLC. Science (Wash DC) 241:353–357

T'ang A, Varley JM, Chakraborty S, Murphree AL, Fung Y-KT 1988 Structural rearrangement of the retinoblastoma gene in human breast cancer. Science (Wash DC) 242:263–266

Yakota J, Akiyama T, Fung Y-KT et al 1988 Altered expression of the retinoblastoma (Rb) gene in small-cell carcinoma of the lung. Oncogene 3:471–475

# Transgenic mice and host cell mutants resistant to transformation as model systems for identifying multiple components in oncogenesis

Harold Varmus

*Departments of Microbiology and Immunology, and Biochemistry and Biophysics, University of California, San Francisco, CA 94143, USA*

*Abstract.* Tumorigenesis appears to be a multistep process involving mutations of conventional, dominantly acting proto-oncogenes, mutations of other genes that may act in a recessive manner, and interactions (or a lack of interactions) between the products of mutant and wild-type genes. Our laboratory is using a few well-established, dominant oncogenes to pose experimental questions that could lead to a better understanding of the more elusive genetic interactions which occur during tumour development. Two such situations are described: (1) We have created a line of transgenic mice that carry the *int-1* proto-oncogene under the control of the enhancer element in the mouse mammary tumour virus long terminal repeat. Such mice express the transgene in mammary glands, salivary glands and male reproductive tract; mammary glands from both male and female animals are grossly hyperplastic, yet tumours arise rarely in the males and sporadically in the females (80% of female mice have one or a few tumours by six months of age). Thus expression of *int-1* in these mice appears to place a large number of mammary cells at risk for secondary events that lead to carcinogenesis, providing a provocative experimental context for identifying such secondary events. (2) We have isolated a rat cell line that lacks most of the characteristics of transformed cells, despite the expression of two wild-type copies of the v-*src* gene of Rous sarcoma virus. This line harbours what appears to be a dominant mutation in an unidentified gene that renders the cell resistant to transformation by v-*src* and several other oncogenes. Isolation of the mutant gene responsible for suppressing transformation in this line should provide new insights into the interactions between oncoproteins and other cellular proteins.

*1989 Genetic analysis of tumour suppression. Wiley, Chichester (Ciba Foundation Symposium 142) p 20–35*

Genetic approaches to carcinogenesis have descended from multiple intellectual lineages and have produced different, even divergent, pictures of the process of tumour formation. Those who have relied upon oncogenic viruses, chemical carcinogens and mutant DNA to induce transformation of cultured cells have generally emphasized the importance of dominantly acting genes in the conversion of cells to a neoplastic phenotype. As a result of rapid advances in molecular genetics, there is now a substantial list of such oncogenes and their wild-type progenitors (proto-oncogenes) and a surprisingly full description of the lesions that render genes oncogenic (for review see Varmus 1984). On the other hand, investigators who have relied upon cells derived from naturally occurring tumours to carry out somatic cell hybridizations, karyology, or analysis of restriction site polymorphisms, have tended to emphasize the importance of recessive mutations in tumorigenesis (Hansen & Cavenee 1987); this school of thought is often supported by analysis of familial cancers (Ponder 1980). However, such mutations are poorly understood at the molecular level, since only one recessively acting locus, the retinoblastoma gene, has been isolated by molecular cloning (see Weinberg, this volume).

It is unlikely that either of these concepts of carcinogenesis is fundamentally wrong; more likely, both dominant and recessive mutations are active at various stages of the multistep process that leads to the development of a cancer cell. Our laboratory has been interested in several experimental strategies that might illuminate the physiological and biochemical interactions proposed to occur among the products of oncogenes, genes that suppress oncogenic activity, and other cellular genes that influence the behaviour of oncoproteins. In this brief essay, I describe two recent results that illustrate the potential for using well-studied, dominantly acting oncogenes to seek other genes involved in carcinogenic mechanisms.

## *int-1* transgenic mice provide a model for multistep tumorigenesis

The *int-1* proto-oncogene was discovered as a frequent target for mouse mammary tumour virus (MMTV) proviral insertion mutations that activate expression of a normally unexpressed gene during mammary carcinogenesis in virus-infected laboratory mice (Nusse & Varmus 1982, Nusse et al 1984). The gene encodes a glycoprotein, of about 40 kDa, that appears to enter the secretory pathway (Brown et al 1987, Papkoff et al 1987). It is usually expressed in only two places: a few cells in the neural tube of mid-gestational embryos and early spermatids in the testis (Shackleford & Varmus 1987, Wilkinson et al 1987). The suggestion that *int-1* might have determining roles during development is reinforced by the finding that its homologue in *Drosophila* is the segment polarity gene, *wingless* (Rijsewijk et al 1987a).

Confirmation of the presumed oncogenic role of this gene in mouse mammary carcinogenesis has come from two sorts of experiments: introduction of transcriptionally active forms of the gene into cultured mammary cells, causing alteration of morphology and growth properties (Brown et al 1986, Rijsewijk et al 1987b, A. Tsukamoto, unpublished), and generation of transgenic mice in which the *int-1* gene is expressed in mammary and salivary glands that subsequently develop adenocarcinomas (Tsukamoto et al 1988).

The *int-1* transgenic mice are especially provocative because they provide a remarkable model for studying multiple steps in carcinogenesis. The transgene was fashioned to resemble a naturally occurring allele from an MMTV-induced tumour, with an MMTV long terminal repeat (LTR) upstream from the four exons of *int-1* in the opposite transcriptional orientation, so that the LTR provides an enhancer for the expression of the gene from its normal promoter. In the single thoroughly studied transgenic mouse line, the transgene is composed of multiple copies of input DNA joined to an unidentified chromosome, transmitted in a simple Mendelian fashion. High levels of transgenic *int-1* RNA are observed in mammary and salivary glands of both male and female animals and in components of the male genital tract (testis, epididymis and vas deferens).

Although the mice show no developmental abnormalities at birth and are sexually active and fertile, they have a number of interesting phenotypic properties. When mammary glands of virgin females and males were removed and examined in whole mount preparations, extensive alveolar–lobular hyperplasia was observed, indicating that expression of *int-1* stimulated marked hyperplasia of mammary epithelium, regardless of the hormonal status of the animals. In the females, the excessive growth has untoward physiological effects, since breeding females are unable to deliver milk to their offspring. (New-born transgenic mice can be nursed successfully by non-transgenic foster mothers.) The hyperplasia is presumed to be an initial step in mammary tumorigenesis because the transgenic mice, unlike their non-transgenic littermates, frequently develop mammary cancers, with an 80% incidence of at least one tumour in female mice by the age of six months. Thus far, only one mammary tumour has arisen in the few male mice that have been kept for eight months. A similar situation seems to exist in the salivary glands, where occasional adenocarcinomas arise in both males and females.

The regular appearance of mammary carcinomas in the female transgenic mice strongly supports the thesis that expression of *int-1* is a causative factor in this disease. Yet it is not a sufficient factor. Large numbers of cells in ten mammary glands express the oncogene and manifest the hyperplastic phenotype, but few cells are converted to frankly malignant cells in up to six months. Thus other apparently uncommon events must conspire with *int-1* in the creation of a cancer cell.

These events could be epigenetic or genetic, dominant or recessive, single or multiple. Cells from transgenic mice should be useful agents for identifying these other elements in the oncogenic pathway. For example, it may be possible to use DNA transfer methods to seek dominantly acting genes that can convert hyperplastic mammary cells into tumorigenic ones; in such experiments, introduction of transiently cultured cells into a cleared mammary fat pad is an assay for the growth potential of the cells. Alternatively, recessive mutations could be sought by karyotyping tumour cells or by looking for the loss of restriction fragment length polymorphism in mice crossed with different strains.

## A cell line, HM, resistant to transformation by the RSV *src* gene

The v-*src* gene of Rous sarcoma virus (RSV) is one of the most intensively studied oncogenes (see Bishop & Varmus 1985, Wyke & Stoker 1987 for reviews). Derived from the highly conserved, normal cellular gene, c-*src*, v-*src* encodes a 60 kDa phosphoprotein that is myristylated, associated with the inner face of the plasma membrane, and endowed with an intrinsic protein kinase activity specific for tyrosine residues. The product of v-*src* differs from the product of its cellular progenitor by virtue of mutations that augment its kinase activity (Sefton & Hunter 1986).

The v-*src* gene driven by a strong transcriptional promoter is among those few powerful oncogenes that can induce the full conversion of a normal cell to a malignant cell in a single step (e.g. Hjelle et al 1988). Despite a lack of requirements for multiple oncogenes, transformation by v-*src* nevertheless depends upon a number of cellular components: targets for its kinase activity, enzymes for chemical modification of the protein (myristyl transferase, phosphatases and other kinases), and factors for localization within the cell. Furthermore, other ill-defined components may be required to make a cell susceptible to transformation, because some cells (such as fibroblasts) are readily transformed, whereas others (such as macrophages) are not. Additional factors are likely to regulate the concentration and activity of c-*src* protein, but it is not known whether the c-*src* kinase phosphorylates a different set of target proteins than does the v-*src* kinase.

A variety of approaches are being taken in our laboratory and others in attempts to define some of these presumed interactions between *src* proteins and other host components. These include intensive site-directed mutagenesis of v-*src* and c-*src* alleles, especially in regions outside the kinase domain; classification of the phosphotyrosine-containing proteins induced in v-*src* transformed cells; purification of *src* proteins, made in large amounts through recombinant DNA technology, for studies of their biochemical and structural properties; and genetic strategies designed to identify cellular genes whose products are likely to interact with *src* proteins. Such strategies take advantage of the existence of v-*src* mutants that transform one type of host cell more efficiently than another (Verderame et al 1989), non-transforming alleles of *src* that might be activated by extragenic mutations, and transformed cells expressing multiple copies of v-*src* such that phenotypic reversion is unlikely to be due to new mutations in the oncogene.

We have recently used this last technique to isolate a cell line that appears to carry a dominant mutation blocking transformation by v-*src* and other potent oncogenes (M.A. Schofield, M. Verderame and H. Varmus, in preparation). A cell line containing three expressed copies of wild-type v-*src* genes (3XSRC) was constructed by infecting an RSV-transformed line of Rat-1 fibroblasts carrying a single, stably expressed provirus (Varmus et al 1981) with two murine leukaemia virus (MLV) vectors carrying v-*src* and different selectable markers. After mutagenesis with ICR 191, a phenotypic revertant was isolated by selective killing of transformed cells and inspection of colonies for morphological reversion.

The revertant cell line (HM) has lost the round, refractile appearance of the 3XSRC parent, it fails to form colonies in soft agar, and it is at least two orders of magnitude less tumorigenic than the parental line in nude mice or syngeneic rats. Nevertheless, the HM line retains two of the three original copies of v-*src* (loss of one of the MLV proviruses may have occurred by chromosomal non-disjunction in the absence of selection). Both v-*src* genes are expressed, as demonstrated by assays for *src* mRNA, *src*-specific tyrosine kinase activity, phosphorylation of a putative target for the kinase, the calpactin I heavy chain (p36), and the pattern of phosphotyrosine-containing proteins in HM cells. Furthermore, both v-*src* genes have been recovered from HM cells in the form of RSV or MLV, and they are competent to induce full transformation when introduced into naive chicken or Rat-1 cells.

These characteristics suggest that the HM cell line is a mutant resistant to transformation by wild-type v-*src*. This conclusion was confirmed by attempts to retransform HM cells with MLV vectors carrying v-*src* and a selectable marker: HM cells expressing the third copy of v-*src* at levels sufficient to transform Rat-1 fibroblasts showed little, if any, alteration in biological properties. In addition, we have been unable to transform HM cells with MLV vectors carrying the v-Ha-*ras* or v-*erbB* oncogenes, suggesting that resistance to transformation is not specific for v-*src* or even for genes encoding tyrosine kinases.

The obvious objective is to isolate and identify the mutant gene responsible for the phenotype of HM cells. Somatic cell hybridizations suggest that the mutation acts in a dominant fashion: fusion of HM cells with Rat-1 cells produces hybrids indistinguishable in appearance from HM cells, and most hybrids formed between HM cells and RSV-transformed Rat-1 cells also resemble HM cells. On the assumption that this interpretation is correct, we are now attempting to re-transform HM cells by superinfection with wild-type MLV, hoping to cause insertion mutations that will inactivate the mutant gene responsible for the HM phenotype.

Strategies similar to that described here for the isolation of HM cells have been used recently for isolation of revertant cells carrying v-Ki-*ras* (Noda et al 1983), v-*fos* (Zarbl et al 1987) and v-*fes* (Haynes & Downing 1988) oncogenes. It will be of considerable interest to compare the affected genes once they have been isolated from the mutant cell lines and to learn how their protein products interact with oncogene products or otherwise participate in neoplastic events.

*Acknowledgements*

Much of the work discussed here was performed by Ann Tsukamoto, Mary Ann Schofield, Mike Verderame, Josh Kaplan and David Morgan, in collaboration with Mike Bishop (UCSF), Rudi Grosschedl (UCSF), Rafe Guzman (UC Berkeley), and Tris Parslow (UCSF). My laboratory is supported by grants from the National Institute of Health, and I am an American Cancer Society Research Professor.

## References

Bishop JM, Varmus HE 1985 Functions and origins of retroviral transforming genes. In: Weiss R, Teich N, Varmus H, Coffin J (eds) RNA Tumor Viruses. Cold Spring Harbor Laboratory, Cold Spring Harbor, NY, vol 1:999–1108

Brown AMC, Wildin RS, Prendergast TJ, Varmus HE 1986 A retrovirus vector expressing the putative mammary oncogene *int*-l causes partial transformation of a mammary epithelial cell line. Cell 46:1001–1009

Brown AMC, Papkoff J, Fung YKT, Shackleford GM, Varmus HE 1987 Identification of protein products encoded by the proto-oncogene *int*-l. Mol Cell Biol 7:3971–3977

Hansen MF, Cavenee WK 1987 Genetics of cancer predisposition. Cancer Res 47:5518–5528

Haynes JR, Downing JR 1988 A recessive cellular mutation in v-*fes* transformed mink cells restores contact inhibition and anchorage-independent growth. Mol Cell Biol, in press

Hjelle B, Liu E, Bishop JM 1988 Oncogene v-*src* transforms and establishes embryonic rodent fibroblasts but not diploid human fibroblasts. Proc Natl Acad Sci USA 85:4355–4360

Noda M, Selinger Z, Scolnick EM, Bassin RH 1983 Flat revertants isolated from Kirsten sarcoma virus-transformed cells are resistant to the action of specific oncogenes. Proc Natl Acad Sci USA 80:5602–5606

Nusse R, Varmus HE 1982 Many tumors induced by the mouse mammary tumor virus contain a provirus integrated in the same region of the host genome. Cell 31:99–109

Nusse R, van Ooyen A, Cox D, Fung YK, Varmus HE 1984 Mode of proviral activation of a putative mammary oncogene (*int*-l) on mouse chromosome 15. Nature (Lond) 307:131–136

Papkoff J, Brown AMC, Varmus HE 1987 The *int*-l proto-oncogene products are glycoproteins that appear to enter the secretory pathway. Mol Cell Biol 7:3978–3984

Ponder PAJ 1980 Genetics and cancer. Biochim Biophys Acta 605:369–410

Rijsewijk F, Schuermann M, Wagenaar E, Parren P, Weigel D, Nusse R 1987a The Drosophila homolog of the mouse mammary oncogene *int*-l is identical to the segment polarity gene *wingless*. Cell 50:649–657

Rijsewijk F, van Deemter L, Wagenaar E, Sonnenberg A, Nusse R 1987b Transfection of the *int*-l mammary oncogene in cuboidal RAC mammary cell line results in morphological transformation and tumorigenicity. EMBO (Eur Mol Biol Organ) J 6:127–131

Sefton BM, Hunter T 1986 Tyrosine kinases, their activators and relatives from c-*src* to v-*src*, or the case of the missing C terminus. Cancer Surv 5:159–173
Shackleford GM, Varmus HE 1987 Expression of the proto-oncogene *int*-1 is restricted to postmeiotic male germ cells and the neural tube of mid-gestational embryos. Cell 50:89–95
Tsukamoto AS, Grosschedl R, Guzman RC, Parslow T, Varmus HE 1988 Expression of the *int-1* gene in transgenic mice is associated with mammary gland hyperplasia and adenocarcinomas in male and female mice. Cell 55:619–625
Varmus HE 1984 Molecular genetics of cellular oncogenes. Annu Rev Genet 18:553–612
Varmus HE, Quintrell N, Wyke J 1981 Revertants of an ASV-transformed rat cell line have lost the complete provirus or sustained mutations in *src*. Virology 108:28–46
Verderame M, Kaplan J, Varmus HE 1989 J. Virol 63:338–348
Weinberg RA 1989 The molecular basis of retinoblastomas. In: Genetic analysis of tumour suppression. Wiley, Chichester (Ciba Found Symp 142) p 99–111
Wilkinson DG, Bailes JA, McMahon AP 1987 Expression of the proto-oncogene *int*-1 is restricted to specific neural cells in the developing mouse embryo. Cell 50:79–88
Wyke JA, Stoker AW 1987 Genetic analysis of the form and function of the viral *src* oncogene product. Biochim Biophys Acta 907:47–69
Zarbl H, Latreille J, Jolicoeur P 1987 Revertants of v-*fos*-transformed fibroblasts have mutations in cellular genes essential for transformation by other oncogenes. Cell 51:357–369

## DISCUSSION

*Klein:* In the ordinary MMTV-induced murine mammary carcinoma, three factors are involved—the virus, hormonal factors and genetic factors. You have shown that the hormonal factors are partially obviated in your system. Have you analysed whether genetic resistance is also obviated in the transgenic animals?

*Varmus:* We are interested in the contribution that the genetic background makes to tumorigenesis. We have carried out the experiments described here in progeny of SJL mice crossed with C57s. Although C57 mice are resistant to MMTV-induced carcinogenesis, our animals are susceptible to the oncogenic effects of the transgene. Further breeding of the transgene into a resistant background may be required to detect any inhibitory effects. The hormonal determinants are not very clear at this point. The frequency of tumours in males seems to be lower than that in females. When we compared breeding females with non-breeding females, the tumour incidence appeared to be higher in the breeding animals; however, the numbers are still small. Philip Leder's group reported that transgenic animals with an activated *neu* oncogene show no difference in tumour incidence between breeding and non-breeding females and no real difference in frequency and time of onset between males and females.

*Wyke:* The tumours induced in transgenic animals did not grow in a pregnancy-dependent manner, whereas those induced by MMTV insertion mutagenesis are initially pregnancy dependent. Do you think this is just a phenomenon related to the level of *int-1* expression?

*Varmus:* It is important to distinguish between two kinds of hormonal effects. One is an hormonal effect on tumour initiation by MMTV; in many strains of mice one can show that the tumour incidence *per se* is strongly pregnancy dependent, even though the tumours themselves don't regress between pregnancies. The A strain mouse is the best example of that. In other strains, like GR, the tumours frequently regress between pregnancies. We see no obvious effect of pregnancy on tumour growth, but, as mentioned earlier, there may be an effect of breeding on the tumour incidence in our *int-1* transgenic animals.

*Klein:* In your somatic hybridization experiment between the revertant and the transformant, the transformed phenotype was largely suppressed, but some foci were transformed. Were they segregants from the chromosomal point of view?

*Varmus:* The karyotypes weren't looked at.

*Klein:* Coming back to the question of revertant resistance to retransformation and comparing your data with Dr Noda's, a relevant question would be whether your revertant is resistant to retransformation by *mos* and by *sis*?

*Varmus:* We have not tried to retransform HM cells with *mos* or *sis*.

*Klein:* Your revertants and the pattern of transformation, as far as you have tested, are consistent with Dr Noda's, aren't they?

*Varmus:* At least one of Dr Noda's cell lines was resistant to *ras* but not *src* transformation. The HM cell seems to be resistant to transformation by both.

*van der Eb:* Should we distinguish between suppression of tumorigenicity and the suppression of transformation, because they are not necessarily always the same phenomenon?

*Varmus:* In the case that we are examining, tumorigenicity and transformation are both suppressed.

*Harris:* Eric, wasn't it you who did the experiment which showed that normal diploid human fibroblasts, if kept in very high serum concentrations, grow in suspension perfectly well?

*Stanbridge:* In soft agar, yes.

*Harris:* This relates to the relationship between transformation *in vitro* and tumorigenicity. You don't have to do anything very sophisticated to 'transform' a diploid fibroblast: all you have to do is increase the serum concentration. So while there is a vague relationship between cells rounding up in culture and their ability to produce tumours, the two events may be separated by many others.

*Stanbridge:* I think one has to be very careful about extrapolating from one system to another. In the past the crime has been to extrapolate from rodent systems to human, you may be committing the crime of extrapolating from human to rodent. Anchorage independence, as far as I understand it, is a very strong correlate of neoplastic transformation in rodent fibroblast cell cultures. Anchorage-independent growth in human diploid fibroblasts is not a good correlate, even of immortalization.

*Tarin:* I have some data on transformation *in vitro* in the mouse mammary gland system. If you take a mouse mammary tumour, this is already tumorigenic and when grown in soft agar gives rise to many clones. If you take each of those individual clones and transplant them back into the mammary fat pad, only about 1% or less will form a new tumour. The fact that the cells will grow in soft agar does not necessarily mean that they are tumorigenic. There is a general correlation between the number of colonies formed in agar and the degree of aggression of a tumour but there is no direct correlation at the level of individual clones. In some tumours, as many as 75% of the colonies form tumours, so it's not a question of the technique. Individual tumours vary in their ability to form clones *in vitro* and in the ability of those clones to grow when transplanted back *in vivo*.

*van der Eb:* In rodent cells there is not always a correlation between tumorigenicity and growth in soft agar. Adenovirus-transformed cells are oncogenic in nude mice or in immunodeficient animals but they grow hardly at all in soft agar or in suspension.

*Noble:* This issue seems to be one of the gaps between people taking the 'bottoms up' approach from the genes and those following the 'top down' approach from the organism. In our studies on gliomas, we were struck by the paradox that gliomas, like most other human tumours, don't grow in tissue culture unless you give them growth factors—either serum or defined mitogens. We looked at the ability of other cells within the tumour, endothelial cells and normal astrocytes, to stimulate glioma cell division. We found that all three members of this three-cell system promote each others' growth. There is a real possibility here of establishing reciprocal feedback loops that may be an intrinsic part of tumour growth *in vivo*, but which are not usually seen in tissue culture. This cannot be detected by the 'bottoms up' approach, which is used to critically define the genetic abnormality. Both are parts of human cancer that one has to consider.

*Sachs:* What we need to clarify is whether there is really dominance and recessiveness in the genetic sense, and to what extent tumorigenesis caused by any agent is a one-step event. If you put the *int-1* gene into a transgenic mouse and the tumours are chromosomally abnormal, as presumably they are, the transgene may have produced the chromosome abnormality, so that even in this case tumorigenesis is a multistep event. Do you think that transfection experiments with NIH-3T3 cells, which are aneuploid, really provide evidence for genetic dominance?

*Varmus:* I don't claim that our transgenes act in a single step; on the contrary, we conclude that secondary events must occur and that our transgenic mice will help define them. I think it's very difficult to defend the argument that tumours arise in a single step, partly for some of the reasons that Al Knudson mentioned: if there are a large number of cells in which secondary events occur frequently, you can't account mathematically for all the biological events that are occurring. Any secondary events could be dominant or recessive in our mice.

*Weinberg:* I think we are dealing with two different classes of tumour suppressor genes. I favour the idea that one needs at least two types of physiological changes in the cell in order to get transformation. The paradigm for me is always *ras* and *myc*. *src* is in the *ras* class and I suggest *Rb* is in the *myc* class.

If one accepts provisionally, independent of the identities of the particular genes involved, that two types of physiological changes are required to get full transformation, one might get reversion from tumorigenicity by two kinds of genetic change. On the one hand, there are downstream targets in the cytoplasm that serve as targets of *ras* and *src* and similarly acting genes. I would argue that the phenomena which Harold Varmus and Dr Noda have uncovered are changes in downstream targets. These changes in cytoplasmic targets represent alleles that are dominant with respect to the action of the cytoplasmic oncogenes and compromise the ability of these genes to elicit their effects.

On the other hand, I think that there are also nuclear changes which are essential for full transformation. These nuclear alterations can be achieved by dominantly acting genes, like the *myc* oncogene. I will speculate that one can elicit similar, although not identical, physiological effects by knocking out recessive genes, like *Rb*. Therefore, by knocking out *Rb* one can achieve a physiological state which resembles that occurring after *myc* activation. Inactivation of *Rb* would yield an apparently recessive tumour phenotype upon cell fusion, whereas the genes that Drs Varmus and Noda talk about would give an apparent dominant phenotype. This has important consequences for the interpretation of cell fusion studies between tumour cells and normal cells. Thus, the ability to revert tumour cells from tumorigenicity by cell fusion doesn't exclude the possibility that such tumour cells may also carry and express a dominantly acting oncogene whose full phenotypic effects cannot be realized because such realization depends on the concomitant co-expression of a second, potentially recessive, type of physiological change within the same cell.

*Varmus:* We are not presuming that the mutant gene in HM cells acts as a *src* suppressor, and we are not suggesting that it in any way affects the *Rb* locus. The mutant gene seems to affect transformation by several oncogenes and may have some general effect on cell growth.

*Weinberg:* I am only trying to establish for the sake of discussion that there are two types of genes which are physiologically and in molecular terms quite different from one another. What you are talking about may be acting at one level, and *Rb* and many other kinds of recessive genes may be acting at another level.

*Stanbridge:* It's interesting that you place *Rb* in the same category as *myc*, I might have put it in another box.

*Sachs:* Again, how do you know that the genes you have referred to are genetically dominant? The cell fusion experiments do not answer this question unless you know the complete karyotype and know that both the relevant alleles are present in the fused cells.

*Weinberg:* If you put *ras* into a normal cell, for example an embryo fibroblast, within hours there are phenotypic changes in that cell. One may not induce tumorigenicity but there indeed are distinct phenotypic changes in the recipient cell. Therefore *ras* is acting dominantly with respect to the normal *ras* proto-oncogene alleles which resided in that cell prior to introduction of the oncogene.

*Sachs:* Is it certain that there are no chromosomal abnormalities in the transfected cells?

*Weinberg:* It happens within hours—that is much too fast for chromosomal changes to have occurred.

*Harris:* It's clear that the words dominant and recessive are being used in two totally different ways. When Bob Weinberg talks about a gene being dominant, he means that if you put it into a cell it does something. That does not mean dominance in any way that would be understood by a geneticist. If, for example, you put the *ras* gene into a cell and produce a phenotypic change, that tells you nothing about dominance or recessivity in a genetic sense. When geneticists use the word dominance they mean that a gene in a diploid organism produces an observable change in the F1 generation and segregates in the F2 generation. In order to see whether a gene inserted into a cell is dominant or not, you have to do a *genetic* experiment. In somatic cells, the only kind of experiment you can do is a somatic cell genetic experiment. If, in a hybrid cell, with proper monitoring of chromosomes, the phenotype initially disappears and subsequently reappears on segregation, the gene is recessive. If the phenotype is present in the hybrid and disappears on segregation, then the gene is dominant. The latter has never been demonstrated for any gene involved in the determination of malignancy.

*Weinberg:* Henry Harris certainly has his right to an opinion about what constitutes genetics, but genetics can be defined both in the sense of organismic genetics and in terms of cell genetics—that is, dominance and recessivity not at the level of the organism but at that of the cell. Accordingly, the introduction of alleles into cells by cell fusion or by molecular means is a perfectly valid way of testing phenotype at the cellular level.

*Cavenee:* The real point is the use of the word recessive to imply loss of function. Recessivity doesn't necessarily mean loss of function, for example, some of Mike Wigler's *ras* mutants of yeast are dominant loss of function mutations (personal communication). The key is whether there is loss of function or not, not whether it's genetically recessive or genetically dominant.

*Harris:* The issue is not whether what one is doing is 'operationally' dominant or recessive, whatever that might mean. The issue is whether the gene or mutation or other event that one is studying is a proximal determinant of the malignant state. For example, we know that if you irradiate a culture with a certain dose of X-rays, you can generate 'transformed' clones in a dish. Nobody thinks that X-irradiation has a highly specific effect in causing transformation; we believe that X-rays act in a more or less random fashion. If the event that you are studying produces 'transformation' at a frequency comparable to that at which it is produced by X-rays, then you are not doing anything very specific. Similarly, if in a transgenic animal you have a gene that is present and active in, say, $10^{14}$ cells and you get one clonal tumour, then the effect is well below what you can do with X-rays and may not be very interesting. On the other hand, if your effect is three or four orders of magnitude better than that produced by X-rays, then you are probably closer to what actually determines the malignant phenotype. So I think what is missing in a lot of these experiments is numerical data.

*Varmus:* We think that the control experiment is to maintain non-transgenic siblings. These don't show mammary hyperplasia and don't have mammary tumours.

*Harris:* There is no doubt that hyperplasia is produced in your experiments, but the question is how many malignant tumours? You suggested that all these hyperplastic cells were clonal in origin.

*Varmus:* I said that the *mammary tumours* in our *int-1* transgenic mice are likely to be clonal, based upon their stochastic appearance. It is difficult to establish clonality rigorously in this experimental setting. We are now infecting the animals with a virus so that the cells are distinguishable by the proviral integration pattern. Then we can ask whether the tumours are clonal or not.

*Harris:* Again, that is not the important issue. If you put a mutated *ras* gene into a cell or inject the mutated *ras* p21 protein, you certainly produce a morphological change: the cell rounds up. (Rounding up is an event of very low specificity: there are thousands of things that cause cells to round up.) The issue is how close is the event that you are observing to generating a malignant tumour. If it is very far away, if several intermediate steps have to occur, then you are dealing with a relatively non-specific perturbation of the cell genome on which selection subsequently operates to produce malignant cells. Anything that increases genetic variability in a system will increase the incidence of tumours - if we accept that tumours have a mutational basis.

*Varmus:* In the *int-1* transgenic mice we have put a lot of cells at risk for secondary tumorigenic events. I don't know how many more events are required, nor what they are. But when we maintain transgenics and non-transgenic siblings under the same conditions, it is the transgenic mice that develop tumours.

*Noble:* Should we be considering a third group of genes that would be analogous to, for example, the *myo*-D gene defined by Weintraub's group (Davis et al 1987), which boost the cell into a differentiated phenotype? In other words, are we looking here at something that bypasses the state of differentiation at which the oncogene-type genes are effective?

*Weinberg:* It is really difficult to know whether the Weintraubian third gene is in any way related to this Weinbergian scheme presented here. It is my prejudice that *myo*-D and similar genes act further downstream. We are dealing here with the central decision of whether or not to differentiate. Once that decision is made, then regulators like *myo*-D will determine which particular differentiation pattern is followed by the cell.

*Noble:* Do you think that Harold (Varmus) has a gene that controls a decision event in terminal commitment?

*Weinberg:* My prejudice is no, he has something earlier and more central.

*Tarin:* You described a mutational event which apparently caused suppression. For me, the issue is what happens in the animal, that is, whether tumours occur in them or not after inoculation. Could you give more details for the incidence of tumours in animals in the various categories that you mentioned? Do you have data on the latent period of tumour induction? All the things that happened in a Petri dish were very interesting, but ultimately we are talking about tumorigenicity in the animal and the only test of that is what happens after the cells have been inoculated into an animal.

*Varmus:* In several experiments, HM cells and subclones were shown to be at least two orders of magnitude less tumorigenic than the parental transformed line when introduced into nude mice or syngeneic rats. On rare occasions, $10^6$ or $10^7$ cells produced small tumours after long latency, similar to results obtained with Rat-1 cells into which no oncogenes were introduced.

There is one potential problem with the HM cell line, namely whether the mutation makes the cell insusceptible to transformation. Do we have a cell that grows poorly and is not robust enough to be transformed by any oncogene? We have tried to introduce into the cell a number of retrovirus vectors that carry a selectable marker plus other oncogenes. v-*src*, of course, doesn't transform the cell. We have tried active Harvey-*ras* and v-*erbB* genes and not succeeded in transforming the cell. We have tried with c-*myc* in a virus vector and do have a few infected cells which grow in soft agar. We also occasionally see spontaneous transformants arising from the HM line; however, it is difficult to determine whether the spontaneous transformants represent loss of the mutation that made the cell non-transformed in the first place. Therefore we are not sure whether we are studying something that has a certain degree of specificity for *src*, whether the cell resists transformation by a variety of oncogenes, or whether the mutation makes the cell less able to grow as vigorously as one would expect for a transformed cell.

*Stanbridge:* That agrees to some extent with Dr Noda's earlier findings

(Noda et al 1983) that there is a finite limit to this resistance; the revertant cell is resistant to transformation by certain oncogenes but is sensitive to others.

*Noda:* I wanted to ask about this specificity to other oncogenes; have you tried *fos* or other tyrosine kinases, such as *abl*?

*Varmus:* No, we haven't tried *abl* or *fos* yet.

*Noda:* Have you looked at the chromosome number of the revertants?

*Varmus:* No.

*Noda:* In this revertant, is the $pp60^{src}$ synthesized and correctly processed, and have you looked at the localization of this protein?

*Varmus:* We haven't looked directly at the localization of the protein because both of the *src* genes in HM cells make proteins that are capable of transforming newly infected cells. Moreover, the pattern of phosphotyrosine-containing proteins is unchanged, which suggests that the *src* proteins are properly localized.

*Noda:* Some people argue that lipidation of the protein might be affected.

*Varmus*: Myristylation mutants of v-*src* are non-transforming, but in this case the *src* genes are not mutant.

*Lydon:* Have you considered the possible overexpression of a phosphatase?

*Varmus:* The pattern of proteins that contain phosphotyrosine is unchanged, which is not what you would have expected if there were an active tyrosine phosphatase.

*Lydon:* This pattern represents all the phosphotyrosine in the cell. However, there may be a small but significant change of a subset of these phosphoproteins.

*Varmus:* In that situation you might expect the *ras* gene to be able to transform cells, since it's generally thought that *ras* acts downstream of *src*.

*Balmain:* It's obvious that there are a number of secondary events involved in generating the tumours. Have you looked for any of these secondary events, for example, *ras* activation, in a transgenic mouse tumour that was not induced by *ras*, or activation of *neu* in the mammary gland tumours?

*Varmus:* We've begun to look in our animals and haven't found anything as yet. One interesting example comes from the work of Anton Berns in Amsterdam (personal communication). He has made transgenic mice that carry the *pim*-1 oncogene. (This oncogene was discovered as a target for an insertion mutation by the murine leukaemia virus.) He made transgenic mice in which the murine leukaemia virus LTR is driving the expression of *pim*-1. These animals have a fairly low tumour incidence, roughly 5%. If one infects those mice at birth with murine leukaemia virus, the tumour incidence goes up to nearly 100% and the onset of tumours occurs appreciably earlier than if the virus were used in a non-transgenic animal. If one looks at the kind of mutation that's caused by the murine leukaemia virus in the transgenic animal, about 85% of the tumours have c-*myc* insertion mutations, and most of the rest have insertion mutations of the N-*myc* gene.

*Knudson:* Do homozygotes show anything different from heterozygotes?

*Varmus:* We are doing that experiment.

*Marshall:* Harold, you mentioned that in the immune complex *src* kinase assay the activity was a bit reduced in the revertant HM cells—do you make anything of that? When you rescue the virus from the HM cells and use this rescued virus to transform rat cells, do you see the reduced level of activity in these transformants?

*Varmus:* We haven't done immune complex assays on the cells that have been transformed by the rescued viruses.

*Marshall:* Then you don't think that reduction in kinase activity is an important issue?

*Varmus:* I don't think so—the levels in HM are still well above the levels that are required for transformation. In addition, no differences were found when the kinase activity was measured with several other assays.

*Marshall:* Have you ever done a mixing experiment, adding cytosol from the revertants to immune complexes from transformed cells and assaying for alterations in kinase activity?

*Varmus:* No.

*Stanbridge:* Concerning the interaction of other proteins with *src*, have you looked at these cells to see if there is anything interesting in terms of protein complexes that include bound *src* protein?

*Varmus:* We haven't looked at other proteins that are in the immune complex with pp60. David Morgan and Josh Kaplan in my group have recently made protein affinity columns using *src* proteins produced by baculovirus. It might be interesting to look at extracts of HM cells in attempts to find proteins that are significant with respect to their affinity for pp60.

*Weinberg:* You have shown that *int-1* can indeed function as an oncogene in mammary epithelia, but what about the implied exclusive relationship between *int-1* and mammary epithelial cells? What about making transgenic mice in which *int-1* would be expressed under the control of a different promoter, so that you don't preordain mammary tumorigenesis?

*Varmus:* We are making a vector containing a histocompatibility gene promoter linked to the *int-1* gene, and that will be put into mice. I would point out that in the existing transgenics we do see tumours in the salivary glands as well as in the mammary glands. However, in the male genital tracts of these transgenic animals there are high levels of *int-1* mRNA, but we see no tumours there. Those tissues may be resistant to the oncogenic properties of *int-1*.

*Knudson:* I referred earlier (Knudson, this volume) to two genetic forms of breast cancer in humans. One of these is associated with soft tissue sarcomas, bone sarcomas, brain tumours and others; these other tumours occur at much lower frequencies. The males sometimes get breast cancer. Have you or anyone else looked at *int-1* loss in breast cancers in humans? Has anyone studied linkage with *int-1* and these familial breast cancers?

*Varmus:* We wouldn't expect loss of *int-1*, but several groups, including ours, have failed to find evidence for expression of *int-1* in human mammary cancers.

*Knudson*: What about linkage?

*Varmus:* It's on chromosome 11q12. There is a tentative linkage between a site of translocation in liposarcomas and the *int-1* locus, identified by Avery Sandberg and colleagues. At least by physical analysis there is no evidence that the *int-1* gene is affected by the translocations.

*Harris:* You said that the hyperplastic mammary epithelium was not functional. Does it produce any milk at all? Could you suckle pups on the male?

*Varmus:* We haven't tried that. The females seem to have milk in the glands, but their pups have no milk in their stomachs. We think that the gland is simply disorganized and unable to deliver milk to the major ducts.

## References

Davis RL, Weintraub H, Lassar AB 1987 Expression of a single transfected cDNA converts fibroblasts to myoblasts. Cell 51:987–1000

Knudson AG 1989 Epidemiology of genetically determined cancer. In: Genetic analysis of tumour suppression. Wiley, Chichester (Ciba Found Symp 142) p 3–19

Noda M, Selinger Z, Scolnick EM, Bassin RH 1983 Flat revertants isolated from Kirsten sarcoma virus-transformed cells are resistant to the action of specific oncogenes. Proc Nat1 Acad Sci USA 80:5602–5605

# Reversion of tumorigenicity in an EBV-converted Burkitt's lymphoma line

George Klein

*Department of Tumor Biology, Karolinska Institutet, Box 60400, S-104 01, Stockholm, Sweden*

*Abstract.* One of five Epstein–Barr virus (EBV)-converted sublines of an EBV-negative Burkitt's lymphoma line (BL-41) was identified as a non-tumorigenic phenotypic revertant with low clonability, comparable to that of an EBV-transformed lymphoblastoid cell line (IARC-171) derived from the same patient. This revertant subline (BL41/95) also showed the most LCL-like phenotype of the five convertants tested. It is suggested that reversion was due to the phenotypic shift of the cell from a 'window' of cell ontogeny that contains virgin B cells and memory B cells to the stage of the activated immunoblast. Constitutive activation of the c-*myc* gene by translocation to an immunoglobulin locus continues to drive proliferation of B cells *in vivo* even after they have undergone a programmed switch to a basically resting (virgin or memory cell) phenotype. An activated immunoblast invariably expresses c-*myc*. It is also suggested that proliferation of the activated immunoblast is regulated by negative host controls that prevent clonal overexpansion and keep the B cell pool constant. The sensitivity of the immunoblast to this control overrides the 'forward-driving' force of both EBV and the activated *myc* gene.

*1989 Genetic analysis of tumour suppression. Wiley, Chichester (Ciba Foundation Symposium 142) p 36–53*

In two recent articles (Klein & Klein 1986, Klein 1987) I have surveyed some of the evidence that activated oncogenes cause malignant (tumorigenic) transformation only if they are expressed at specific stages ('windows') of differentiation. If this notion is correct, it follows that experimentally induced shifts in differentiation may push the tumour cell into an 'inactive window'. Continued expression of the original transforming oncogene would then no longer suffice to drive cell division forward.

These expectations are fulfilled by the successful induction of terminal differentiation in teratomas, erythroleukaemias, myeloid leukaemias or neuroblastomas and by the release of the differentiation block induced by a temperature-sensitive v-*src* or v-*erbB* oncogene when the transformed cell is shifted to a non-permissive temperature (reviewed in Klein & Klein 1986). They are also consistent with the imposition of the full differentiation programme on a highly malignant cell by hybridization with a normal keratinocyte or fibroblast partner (Stanbridge 1987, Harris, this volume).

B lymphocytes develop and mature according to a strictly regulated programme that is well known in outline but only incompletely known in detail (Nilsson & Klein 1982, Godal & Funderud 1982). The situation is complicated by the interdigitation of two partly linked, but distinct, processes, maturation and activation. Activation can take place at different levels of differentiation. Both the virgin B lymphocyte and its clonally expanded but quiescent counterpart, the memory cell, are heterogenous with regard to the level of maturation, as judged by the production of surface or secretory immunoglobulin, the development of a Golgi apparatus and the appearance of other differentiation-associated markers. In other words, activated B cells may proliferate and virgin or memory B cells may rest in G0 at different maturation stages. They are distinguishable by the presence or absence of activation markers.

Germinal centre memory cells express the B cell markers CALLA and BLA but no activation markers (Gregory et al 1987). Their phenotype closely resembles that of BL cells (Ehlin-Henriksson et al 1987, Rowe et al 1985). When exposed to Epstein–Barr virus (EBV), resting B cells lose CALLA and BLA and express a set of activation markers. Established lymphoblastoid cell lines (LCLs) have the same phenotype as activated immunoblasts.

Table 1 summarizes the main phenotypic properties of LCL and BL lines. For BL lines, the table refers to the typical or type I phenotype, according to the classification of Rowe et al (1985), which is characteristic of the BL cell *in vivo*, including both the highly endemic, largely EBV-carrying, and the sporadic, largely EBV-negative, forms. The EBV-negative BL lines remain phenotypically stable during prolonged *in vitro* propagation, whereas EBV-positive lines tend to drift towards a more LCL-like (type II or type III) phenotype.

LCL and BL lines differ in their EBV antigen expression. Seven proteins are regularly expressed in the LCLs: six are nuclear (designated EBNA-1–6), whereas the seventh is called latent membrane protein (for review see Dillner & Kallin 1988). Type I BL lines express only one of the seven proteins, EBNA-1, which is known to bind to the replication origin (*oriP*) of the viral genome (Yates et al 1985). This binding appears to be required for the maintenance of the viral episomes and it also triggers viral enhancer activity. During the phenotypic drift of EBV-positive BL cells towards a more LCL-like phenotype, EBNA-2–6 and latent membrane protein expression is switched on.

BL lines and LCLs also differ in their tumorigenicity and in their clonability in agarose (Nilsson et al 1977, Giovanella et al 1979). BL lines are highly clonable and tumorigenic in immunodeficient mice. Diploid LCLs of relatively recent origin are non-clonogenic and non-tumorigenic in adult nude mice. They grow temporarily in new-born nude mice, and can grow progres-

**TABLE 1 Main phenotypic characteristics of lymphoblastoid (LCL) and Burkitt's lymphoma (BL) lines**

| | *Growth pattern in vitro* | *Presence of Ig/myc trans-location* | *Agarose clon-ability* | *CALLA BLA* | *Acti-vation markers* | *EBNA-1* | *EBNA-2* | *EBNA-3–6* | *Latent membrane protein* | *HLA class I expression (e.g. A11)* |
|---|---|---|---|---|---|---|---|---|---|---|
| LCL | Clumps of immuno-blasts | – | low or nil | – | +++ | + | + | + | + | high |
| BL lines (type I) | single cells | + | high | + | – | + | – | – | – | low or nil |

sively in the brains of nude but not of immunocompetent mice. After long-term *in vitro* culturing, LCLs tend to become aneuploid, with numerical rather than structural chromosomal changes (mainly trisomies). Subsequently, they may show increased clonability and tumorigenicity.

The difference in clonability in agarose is an indicator of the autonomous growth potential of the two prototype cells. The tumorigenicity experiments in nude mice show an analogous difference. This is not an artifact of the heterograft system due to differences in natural killer cell sensitivity, because LCLs are equally sensitive to lysis by natural killer cells as the BL lines of the same donors, or less sensitive (Torsteinsdottir et al 1984).

BL cells are resistant to the lytic effect of EBV-specific cytotoxic T cells (CTL) generated by stimulating normal T cells with autologous EBV-transformed B cell lines (Rooney et al 1984, Torsteinsdottir et al 1986); LCLs from the same patients are killed with high efficiency. Three phenotypic differences between the two cell types may, individually or collectively, account for this resistance of the BL cells. BL cells express some MHC class I polymorphic determinants (HLA-A11 in particular) at a much lower level than LCLs derived from the same donor (Torsteinsdottir et al 1986, 1988a, Masucci et al 1987) (Table 1). Certain leucocyte adhesion molecules, required for cell-mediated cytotoxic reactions, are also expressed at a lower level in BL cells (Patarroyo et al 1988, Gregory et al 1988a). Thirdly, type I BL cells do not express EBNA-2–6 and latent membrane protein; EBNA-2 and latent membrane protein, or their processed peptide products, are currently regarded as potential targets of immune surveillance.

The non-tumorigenicity of LCLs is thus established. We have also shown that EBV-negative BL cells can be induced to change to a more LCL-like phenotype by EBV (see below). Can they also lose their tumorigenicity? This question has been studied with Sigurbjörg Torsteinsdottir, Maria Grazia Masucci, Barbro Ehlin-Henriksson and Eva Klein (Torsteinsdottir et al 1988b).

As our main target cell, we selected BL41/95, a subline of the EBV-negative BL-41 lymphoma converted by the B95–8 substrain of EBV. Of the five convertants tested, BL41/95 had shown the most change to an LCL-like phenotype (Torsteinsdottir et al 1986, Masucci et al 1987). We compared its clonability and tumorigenicity with those of the other four BL-41 lines converted by EBV. Our results indicate that the phenotypic switch is accompanied by a loss of tumorigenic behaviour.

## Results

### *Surface marker expression and sensitivity to CTL*

Expression of activation and differentiation markers, HLA class I antigen and EBV latent membrane protein by the BL-41 lines and the LCL derived from

**TABLE 2 Characteristics of the BL-41 family of cell lines**

| | | *BL-41* | *BL-41/95* | *E95A–BL-41* | *E95B–BL-41* | *E95C–BL-41* | *EHRA–BL-41* | *IARC–171* |
|---|---|---|---|---|---|---|---|---|
| B cell markers[a] (% positive cells) | LB-1 | 6±1 | 52±10 | 46±5 | 40±5 | 13±8 | 22±6 | 100 |
| | CD21 | 7±3 | 72±5 | 29±6 | 37±9 | 31±4 | 11±5 | 35±5 |
| | CD23 | 0 | 47±9 | 22±7 | 34±7 | 31±8 | 26±6 | 84±4 |
| | CD10 | 100 | 98±2 | 85±6 | 93±1 | 92±4 | 96±2 | 3±2 |
| | BLA | 64±10 | 8±5 | 45±12 | 27±8 | 5±1 | 11±5 | 6±4 |
| HLA Class I expression[b] (% mean fluorescence: volume ratio) | A,B,C | 42±7 | 45±11 | 47±11 | 40±2 | 44±8 | 42±10 | 41±14 |
| | A11 | 10±6 | 29±5 | 15±5 | 13±8 | 14±2 | 13±2 | 30±3 |
| | A32 | 5±3 | 16±3 | 13±2 | 14±1 | 14±3 | 13±1 | 17±3 |
| | A32 | 18±4 | 28±3 | 27±3 | 27±2 | 26±4 | 24±3 | 35±3 |
| | Bw6 | 26±2 | 35±2 | 33±1 | 36±2 | 34±4 | 30±2 | 36±4 |
| Sensitivity to anti-HLA CTL lysis[c] (% specific lysis) | anti-A11 | 0 | 37±12 | 1±1 | 2±1 | 0 | 3±2 | 71±9 |
| | anti-B35 | 14±6 | 51±7 | 30±5 | nd | 25±8 | nd | 72±5 |
| | anti-Cw4 | 51±7 | 78±9 | 61±1 | nd | 62±7 | 69±5 | 80±7 |
| EBV latent membrane protein expression[d] | | 0 | 325 | ≤2 | ≤2 | ≤2 | ≤2 | 3250 |

[a] Reactivity with the monoclonal antibodies was measured by indirect immunofluorescence. Mean±SE of 3–8 experiments. [b] The mean fluorescence:volume ratio was calculated as described in Torsteindottir et al (1988a). Mean±SE of 3–6 experiments. [c] The % specific lysis in 4 hour $^{51}$Cr-release assays performed at 10:1 effector:target ratios. Mean±SE of 3–5 experiments. [d] The reciprocal of the cell extract dilution at which EBV latent membrane protein was detectable.

the same donor (designated IARC-171) is summarized in Table 2, together with the sensitivity of these lines to HLA-specific CTL clones. As reported elsewhere (Ehlin-Henriksson et al 1987, Avila-Carino et al 1987, Torsteinsdottir et al 1988a), BL-41 expressed the B cell markers CD10 (CALLA) and BLA, but none of the three activation markers tested. All EBV-converted BL-41 sublines expressed the activation markers LB-1, CD23 and, to a lesser extent, CD21. They maintained a high level of CD10 expression, but were less positive for BLA than was the BL-41 parental line.

All lines reacted equally well with the anti-HLA-A,B,C monoclonal antibody W6/32. The polymorphic determinant HLA-A32 and the public specificity defined by Bw6 were expressed at a lower level on BL-41 than on the convertants. A11 was detected minimally on BL-41 or not at all (Masucci et al 1987). IARC-171 and BL41/95 showed marked expression of HLA-A11, whereas the other EBV-converted BL-41 sublines tested expressed A11 at very low levels.

The sensitivity of the lines to an HLA-A11-specific CTL clone paralleled the results on antibody binding. IARC-171 was highly sensitive, whereas BL-41 and its EBV-converted sublines were totally resistant, except BL41/95 which showed intermediate sensitivity. BL-41 was also relatively resistant to an HLA-B35-specific CTL clone. Of the three convertant lines tested, BL41/95 showed most increase in sensitivity to lysis by this clone. Sensitivity to a Cw4-specific CTL clone varied similarly, although it lysed 50% of BL-41 cells.

All BL-41 convertants except EHRA-BL-41 expressed all six antigens of the EBNA family (Torsteinsdottir et al 1988b). The P3HR1-converted EHRA-BL-41 line did not express EBNA-2. Only BL41/95 expressed the virus-encoded latent membrane protein (Masucci et al 1987).

### *Cloning efficiency in soft agarose*

The cloning efficiency of the BL-41 parent varied between 52 and 82% in

**TABLE 3 Cloning efficiency of BL-41 and its EBV-converted sublines**

| *Cell line* | *Cloning efficiency in single experiments (mean ± SD)*[a] | | | | | |
|---|---|---|---|---|---|---|
| BL-41 | 67±8 | 52±14 | — | 82±7 | 68±19 | 68±10 |
| BL41/95 | 3±1 | 14±1 | — | — | 20±7 | 23±4 |
| E95A-BL-41 | 95±8 | 92±8 | — | 87±9 | — | 90±3 |
| E95B-BL-41 | 89±10 | 54±16 | — | — | — | — |
| E95C-BL-41 | 75±14 | 45±12 | — | — | — | — |
| EHRA-BL-41 | — | — | 46±2 | 81±10 | — | — |
| IARC-171[b] | 3±2 | — | — | — | 2±1 | — |
| Raji[c] | 83±6 | 93±5 | 69±4 | 68±1 | 89±15 | 87±14 |

[a] Mean ± SD of counted colonies in three agar plates seeded with 100 cells each.
[b] IARC-171, lymphoblastoid cell line derived from the same donor as BL-41.
[c] Raji, an EBV-carrying BL line (positive control).

different experiments (Table 3): that of the positive control, the EBV-carrying BL line Raji, was 68–93%. IARC-171 had a very low cloning efficiency (2–3%), as expected. Four of the five EBV-converted BL-41 lines retained a high cloning efficiency, whereas BL-41/95 showed a significant reduction, down to 3–23%.

### *Tumorigenicity in immunosuppressed mice*

The BL-41 lines and IARC-171 were inoculated subcutaneously into immunosuppressed mice. The cumulative take incidence after two weeks is shown in Table 4. BL-41 grew in 56, 31 and 39% of the mice inoculated with 20, 10 and 5 × $10^6$ cells, respectively. IARC-171 did not grow at all, apart from a single take among 18 mice inoculated with 20 × $10^6$ cells. Four convertants had a considerably higher take incidence than the original BL-41 line. In contrast, BL41/95 had a significantly reduced take incidence: inoculations of 5 × $10^6$ and 10 × $10^6$ cells did not grow at all, and one of 20 × $10^6$ cells grew in only 20% of the mice. The majority (50–80%) of tumours regressed after 3–6 weeks of growth. All BL41/95-derived tumours and the single LCL-derived tumours regressed within three weeks.

**TABLE 4 Tumorigenicity of BL-41 and its EBV-converted sublines in immunosuppressed mice**

| | *Inoculum dose and take incidence*[a] | | | | | |
|---|---|---|---|---|---|---|
| *Cell line* | *5 × $10^6$* | % | *10 × $10^6$* | % | *20 × $10^6$* | % |
| BL-41 | 7/18 | 39 | 5/16 | 31 | 44/78 | 56 |
| BL41/95 | 0/59 | 0 | 0/25 | 0 | 15/73 | 20 |
| E95A-BL-41 | 28/37 | 76 | 10/10 | 100 | 60/85 | 70 |
| E95B-BL-41 | 6/6 | 100 | — | — | 20/22 | 91 |
| E95C-BL-41 | — | — | — | — | 38/46 | 83 |
| EHRA-BL-41 | — | — | 5/12 | 42 | 10/11 | 91 |
| IARC-171 | 0/18 | 0 | — | — | 1/18 | 5 |

[a] Number of mice with growing tumours after two weeks/total number inoculated and percentage take incidence.

### *Steady-state levels of c-myc mRNA*

Total cellular RNA from exponentially growing cells was separated on agarose gels and hybridized to a probe recognizing the third exon of human c-*myc* and to the control GAPDH (glyceraldehyde-3-phosphate dehydrogenase) probe. The intensity of the specific bands was quantified by densitometry and a specific c-*myc*:GAPDH ratio was calculated for each line. IARC-171, and BL-41 and its EBV-converted sublines, were tested in parallel. IARC-171 showed the lowest c-*myc*:GAPDH ratio and BL-41 the highest (data not shown). The level of c-*myc* mRNA was 4–5 times higher in the

BL cells than in the LCL. The c-*myc*:GAPDH ratios of all converted BL-41 sublines were intermediate between the high ratio for BL-41 and the low one for LCL. There was no systematic difference between BL41/95 and the tumorigenic EBV convertants.

## Discussion

These results show that BL cells can become non-tumorigenic, in spite of the continued presence of the IgH/*myc* translocation chromosome and the same high level of c-*myc* transcription as in the parallel EBV-converted, highly tumorigenic sublines. The revertant BL41/95 was unique among the five EBV-convertants; it showed the most extensive up-regulation of the B cell activation markers and it was the only convertant in the BL-41 series that was induced to express HLA-A11 and EBV latent membrane protein, which were highly expressed in the corresponding LCL (IARC-171).

It may be surmised that the shift of phenotype has moved BL41/95 out of the 'tumorigenic transformation-sensitive window' — that is, the narrow phenotypic band where the cell can be driven by an activated *myc* gene to grow progressively *in vivo*. It is interesting to contrast this with the induction of differentiation in other transformed or tumour cell systems (for review see Klein & Klein 1986). Certain myeloid leukaemias, erythroleukaemias, teratomas and neuroblastomas can be induced to differentiate by physiological and/or chemical signals. In some cases, this was shown to be correlated with the down-regulation of a certain oncogene, c-*myc* or N-*myc* in particular, that was believed to have contributed to tumour progression.

In B lymphocytes, differentiation is compatible with continued cell proliferation. Malignant tumours can develop from the presumptive terminal cell, the plasmacyte (Klein & Klein 1986). There is no convincing evidence that plasma cells ever lose their ability to divide, as do terminally differentiated neurons, granulocytes or macrophages.

In our present system, reversion to non-tumorigenicity is related to the appearance of activation, rather than differentiation markers. The possible reasons for this reversion may be discussed in two contexts: Burkitt's lymphoma and immunoblast activation.

Different scenarios have been suggested for Burkitt's lymphoma (BL) (reviewed in Klein & Klein 1986). A key question concerns the stage of B cell differentiation when the characteristic Ig/*myc* translocation is most likely to occur. Current views are that the 'translocation accident' takes place at the time of VDJ arrangements in the endemic form of BL and at the time of the heavy chain switch recombination in sporadic BL. These conclusions were deduced from the most frequent sites of *myc* rearrangement in the typical 8;14 translocation. Some of our recent findings question these conclusions, however. With Ender Altiok and Ingemar Ernberg, we have recently found that four of six independently maintained sublines of the same EBV-transformed pro-B cell line (Katamine et al 1984) developed $14q^+$ markers, generated by the translocation of terminal chromosome pieces from different chromosomes (not chromosome 8) to the *Smu* region of the IgH locus. The origin of the transposed piece was traced to chromosome 6 in one subline in an earlier collaborative study (Otsu et al 1987). The translocation-carrying cell has overgrown the normal diploid cell of origin in each of the four lines. This suggests that the *Smu* region may provide a hot spot for illegitimate recombination in the pro-B cell, before the initiation of normal DNA rearrangement. The translocation of genes that favour *in vitro* growth could lead to the competitive replacement of the original diploid line, as observed in the four sublines. The corresponding transposition of c-*myc* might generate an equivalent growth advantage *in vivo*. If so, the translocation-carrying cell would still be able to proceed with its normal immunoglobulin gene rearrangement, since all Burkitt's lymphoma and derived lines make a heavy chain (encoded by the non-translocated chromosome), and most of them also make a light chain.

We have previously suggested (Klein & Klein 1985) that constitutive activation of the *myc* gene would prevent the B cell from resting in G0 at the stage when it is programme to do so, i.e. at the virgin and/or the memory B cell stages. The accidental misplacement of c-*myc* under the control of an Ig gene that is constitutively expressed in all B cells would inhibit the normal down-regulation of the gene. This would prevent the cell from leaving the cycling compartment, perhaps because of a direct or indirect stimulatory effect of the *myc* protein on DNA synthesis (Classon et al 1987).

The EBV- or mitogen-activated immunoblast is phenotypically a very different cell. At least one of its activation markers (CD23) is a growth factor receptor (Gordon et al 1986). This is reminiscent of the appearance of the interleukin 2 receptor on HTLV-1 virus-infected T cells (Gootenberg et al 1981). The induction of specific growth factor receptors may provide an advantage for viral survival.

After being triggered by the appropriate activating complex, the immunoblast expands clonally. The size of the B cell pool is constant and cannot be increased by hyperimmunization; it follows that strict feedback controls must prevent the overproliferation of each B cell clone (Osmond 1986). Prolonged antigenic stimulation could otherwise readily lead to progressive lymphoproliferative disease. Both the virgin and the memory B cell must be insensitive to this feedback. This should normally pose no problems, since these cells are securely positioned in the resting stage. Their activation requires a complex set of signals, emitted by macrophages, dendritic cells and T cells after antigen processing and presentation on the appropriate MHC molecule. There seems to be no need for extra safeguards to stop resting B cells from proliferating. After constitutive activation of the *myc* gene, the inhibitory feedback signals that prevent the over-expansion of the activated blasts would therefore fail to influence the growth of the translocation-carrying cell, as long as it maintains the phenotypic traits of the 'resting' virgin and/or memory cell. This would explain the narrow 'window' associated with the oncogenic effect of the *myc*/Ig transposition.

It follows that the EBV-transformed immunoblast is under the influence of feedback signals that control the size of the B cell pool. However, the EBV genome does have the potential to drive cell proliferation, as proved by the continuous growth of the EBV-transformed LCLs *in vitro*.

Viral survival is favoured by host survival. If the same host controls are responsible for regulating the proliferation of physiologically triggered and EBV-activated immunoblasts, the latter can constitute only a small, relatively constant part of the total B cell pool. This is supported by evidence suggesting an equilibrium between latent EBV-carrying cells and their host in normal individuals (Nilsson et al 1971, Lewin et al 1987). It is also in line with the non-tumorigenicity and low clonability in agarose of the EBV LCL (Nilsson et al 1977, Giovanella et al 1979). While it is not clear what this clonability actually measures, its correlation with tumorigenic behaviour agrees with experience from other neoplastic systems.

Severely immunodeficient hosts, such as transplant patients or children with the X-linked lymphoproliferative syndrome, are often unable to control the growth of EBV-carrying blasts (for review see Purtilo & Klein 1981). This is usually interpreted to mean that EBV-carrying cells are under T cell-mediated immunological control in the normal host. T cells can act in ways other than by straightforward CTL-mediated lysis, however. They may also play an important role in controlling the B cell pool size by feedback signals. The reversion of the tumorigenic phenotype in the BL41/95 cell illustrates that the extensive induction of activation markers in a BL cell by EBV conversion may occasionally be paralleled by the 'reversion' of the cell to an LCL-like responsiveness to growth control. The inability of the constitutively activated, translocated *myc* gene to override this phenotypic trait may reflect the fact that down-regulation of c-*myc* is not normally a key element in the control of immunoblast proliferation. EBV-transformed LCLs express c-*myc* at levels similar to or slightly lower than those of BL cells. If normal blasts are expanding in response to the appropriate antigenic and lymphokine signals, but under the control of a feedback mechanism that prevents them from exceeding a given pool size, it is reasonable to assume that this control acts as a brake, in contrast to the complete stop signal that brings the virgin and the memory B cell to a standstill in G0 in parallel with a major phenotypic switch. The interpretation that the effect of the translocated *myc* gene can be overridden by the LCL phenotype is also consistent with the fact that the LCL phenotype dominates in LCL/BL cell hybrids (Ling & Lowe 1987, B. Ehlin-Henriksson et al, to be published).

It may appear paradoxical that EBV, a highly transforming and potentially oncogenic virus, can induce phenotypic changes that normalize a Burkitt's lymphoma cell. The effects of the virus are complex, however. They include not only the forward-driving force, manifested by *in vitro* immortalization, but also the ability to induce a major phenotypic switch in B cells, manifested as immunoblast activation. If the latter switch is sufficiently radical, it may lead to the 'dominant reversion' exemplified by BL41/95. The lack of similar reversion in other EBV-converted BL-41 lines indicates that the 'radical switch' is not easy and not regular. This fits with the fact that BL cells are relatively, although not completely, refractory to activation by EBV or mitogens. Gregory et al have shown (1988b) that EBV infection of the closest normal counterparts of the BL cell — the CALLA- and BLA-positive, activation marker-negative germinal centre B cell — leads to the prompt disappearance of CALLA and BLA and the appearance of the activation markers. EBV infection of EBV-negative BL cells with the same phenotype does not lead to a comparably prompt and radical switch. Rather, EBV-converted BL lines show a slow and gradual appearance of activation markers, in parallel with the initial maintenance of CALLA and BLA, as in the present case. The extent of this change varies between different target cells and also between individual convertants (Ehlin-Henriksson et al 1987, Calender et al 1987) but its eventual dominance in the long-established EBV-carrying BL lines suggests that the LCL-like variants have a selective advantage *in vitro*.

*Acknowledgements*

This investigation was supported by PHS grant number 5 RO1 CA14054–15 from the National Cancer Institute, DHHS and a grant from the Swedish Cancer Society.

## References

Avila-Carino J, Torsteinsdottir S, Ehlin-Henriksson B et al 1987 Paired Epstein-Barr virus (EBV)-negative and EBV-converted Burkitt lymphoma lines: stimulatory capacity in allogeneic mixed lymphocyte cultures. Int J Cancer 40:691–697

Calender A, Billaud M, Aubry J-P, Banchereau J, Vuillaume M, Lenoir G M 1987 Epstein-Barr virus (EBV) induces expression of B-cell activation markers on in vitro infection of EBV-negative B-lymphoma cells. Proc Natl Acad Sci USA 84:8060–8064

Classon M, Henriksson M, Sumegi J, Klein G, Hammarskjöld M-L 1987 Elevated c-myc expression facilitates the replication of SV40 DNA in human lymphoma cells. Nature (Lond) 330:272–274

Dillner J, Kallin B 1988 The Epstein-Barr virus proteins. Adv Cancer Res 50:95–158

Ehlin-Henriksson B, Manneborg-Sandlund A, Klein G 1987 Expression of B-cell-specific markers in different Burkitt lymphoma subgroups. Int J Cancer 39:211–218

Giovanella B, Nilsson K, Zech L et al 1979 Growth of diploid, Epstein-Barr virus-carrying human lymphoblastoid cell lines heterotransplanted into nude mice under immunologically privileged conditions. Int J Cancer 24:103–113

Godal T, Funderud S 1982 Human B-cell neoplasm in relation to normal B-cell differentiation and maturation processes. In: Klein G and Weinhouse S (eds) Adv Cancer Res. Academic Press, New York, vol. 36:211–255

Gootenberg J E, Ruscetti F W, Mier J W, Gazdar A F, Gallo R C 1981 Human cutaneous T-cell lyphoma and leukemia cell lines produce and respond to T-cell growth factor. J Exp med 154:1403–1418

Gordon J, Webb A J, Walker L, Guy G R, Rowe M 1986 Evidence for an association between CD23 and the receptor for a low molecular weight B cell growth factor. Eur J Immunol 16:1627–1630

Gregory C, Tursz T, Edwards C F et al 1987 Identification of a subset of normal B cells with a Burkitt's lymphoma (BL)-like phenotype. J Immunol 139:313–318

Gregory C D, Murray R J, Edwards C F, Rickinson A B 1988a Down-regulation of cell adhesion molecules LFA-3 and ICAM-1 in Epstein–Barr virus-positive Burkitt's lymphoma underlies tumour cell escape from virus-specific T cell surveillance. J Exp Med 167:1811–1824

Gregory C D, Edwards C F, Milner A et al 1988b Isolation of a normal B cell subset with a Burkitt's like phenotype and transformation in vitro with Epstein–Barr virus. Int J Cancer 42:213–220

Harris H 1989 The biology of tumour suppression. In: Genetic analysis of tumour suppression. Wiley, Chichester (Ciba Found Symp 142) p 199–213

Katamine S, Otsu M, Tada K et al 1984 Epstein-Barr virus transforms precursor B cells even before immunoglobulin gene rearrangements. Nature (Lond) 309:369–372

Klein G 1987 The approaching era of the tumor suppressor genes. Science (Wash DC) 238:1539–1545

Klein G, Klein E 1985 Myc/Ig juxtaposition by chromosomal translocations: some new insights, puzzles and paradoxes. Immunol Today 6:208–215

Klein G, Klein E 1986 Conditioned tumorigenicity of activated oncogenes. Cancer Res 46:3211–3224

Lewin N, Åman P, Masucci M G et al 1987 Characterization of EBV-carrying B-cell populations in healthy seropositive individuals with regard to density, release of transforming virus and spontaneous outgrowth. Int J Cancer 39:472–476

Ling N R, Lowe J 1987 Regulation of B cell antigens on immunoglobulin-secreting cells. Experiments with human-human and human-mouse hybrid cells (abstract). British Society for Immunology, Autumn Meeting p. 91

Masucci M G, Torsteinsdottir S, Colombani J, Brautbar C, Klein E, Klein G 1987 Down-regulation of class I HLA antigens and of the Epstein-Barr virus encoded latent membrane protein in Burkitt lymphoma lines. Proc Natl Acad Sci USA 84:4567–4571

Nilsson K, Klein G 1982 Phenotypic and cytogenetic characteristics of human B-lymphoid cell lines and their relevance for the etiology of Burkitt's lyphoma. In: Klein G, Weinhouse S (eds) Adv Cancer Res. Academic Press, New York, vol 37:319–380

Nilsson K, Klein G, Henle W, Henle G 1971 The establishment of lymphoblastoid lines from adult and fetal human lymphoid tissue and its dependence on EBV. Int J Cancer 8:443–450

Nilsson K, Giovanella B C, Stehlin J S, Klein G 1977 Tumorigenicity of human hematopoietic cell lines in athymic nude mice. Int J Cancer 19:337–344

Osmond D G 1986 Population dynamics of bone marrow B lymphocytes. Immunol Rev 96:103–124

Otsu M, Katamine S, Uno M et al 1987 Molecular characterization of novel reciprocal translocation t(6;14) in an Epstein-Barr virus-transformed B cell precursor. Mol Cell Biol 7:708–717

Patarroyo M, Prieto J, Ernberg I, Gahmberg C G 1988 Absence, or low expression, of leukocyte adhesion molecules CD11 and CD18 on Burkitt lyphoma cells. Int J Cancer 41:901–907
Purtilo D T, Klein G 1981 Introduction to Epstein-Barr virus and lymphoproliferative diseases in immunodeficient individuals. Cancer Res 41:4209–4211
Rooney C M, Rickinson A B, Moss D J, Lenoir G M, Epstein M A 1984 Paired Epstein-Barr virus-carrying lymphoma and lymphoblastoid cell lines from Burkitt's lymphoma patients: comparative sensitivity to non-specific and to allo-specific cytotoxic responses, in vitro. Int J Cancer 34:339–348
Rowe M, Rooney C M, Rickinson A B et al 1985 Distinctions between endemic and sporadic forms of Epstein-Barr virus-positive Burkitt's lymphoma. Int J Cancer 35:435–441
Stanbridge E J 1987 Genetic regulation of tumorigenic expression in somatic cell hybrids. In: Klein G (ed) Adv in Viral Oncol. Raven Press, New York, vol 6:83–101
Torsteinsdottir S, Masucci M G, Lenoir G, Klein G, Klein E 1984 Natural killer cell sensitivity of human lymphoid lines of B-cell origin does not correlate with tumorigenicity or with the expression of certain differentiation markers. Cell Immunol 86:278–286
Torsteinsdottir S, Masucci M G, Ehlin-Henriksson B et al 1986 Differentiation-dependent sensitivity of human B-cell-derived lines to major histocompatibility complex-restricted T-cell cytotoxicity. Proc Natl Acad Sci USA 83:5620–5624
Torsteinsdottir S, Brautbar C, Klein G, Klein E, Masucci M G 1988a Differential expression of HLA antigens on human B-cell lines of normal and malignant origin: a consequence of immune surveillance or a phenotypic vestige of the progenitor cells? Int J Cancer 41:913–919
Torsteinsdottir S, Andersson M, Avila-Carino J et al 1988b Reversion of tumorigenicity and decreased agarose clonability after EBV conversion of an IgH/myc translocation carrying BL line. Int J Cancer, in press
Yates J L, Warren N, Sugden B 1985 Stable replication of plasmids derived from Epstein–Barr virus in various mammalian cells. Nature (Lond) 313:812–815

## DISCUSSION

*Sachs:* You obtained revertants from the virus-infected cells but not from the non-infected cells. Were the chromosomes of the revertants different from those of the parental line?

*Klein:* No.

*Stanbridge:* In your studies of Burkitt's lymphoma, were the assays for tumorigenicity done at sub-cutaneous sites or intra-cerebrally?

*Klein:* They were all done sub-cutaneously.

*Stanbridge:* Have you tried injecting them intra-cerebrally?

*Klein:* No, because in earlier studies even the LCLs grew progressively in the brain.

*zur Hausen:* We have some evidence that the development of Burkitt's lymphoma occurs beyond the level of the activated oncogene expression. Jürgen Wolf in our laboratory established a number of hybrids between Burkitt's lymphoma cells and a lymphoblastoid line obtained from the same patient. All of these hybrids, which have been carefully analysed, contain the translocated *myc*; they all retain the lymphoblastoid phenotype; none grows invasively after sub-cutaneous inoculation into nude mice, (usually, inoculation leads to initial tumour formation, but after a short period the tumours necroticize and regress completely). So we feel that there is some kind of negative control functioning in these hybrids in spite of the presence of the modified oncogene.

Concerning the cell line, did you investigate whether EBV conversion of other EBV-negative tumours leads to the same event?

*Klein:* We haven't looked at many, but we don't have this extreme phenotypic switch in any of them. I was delighted to hear about your hybrids, because your findings agree completely with our hypothesis. Have you examined the expression of the EBV antigens, particularly EBNA-1 and EBNA-2? I would expect to see an up-regulation.

*zur Hausen:* We have not looked at that yet. We would also expect that, because the phenotype is absolutely identical to that of the lymphoblastoid line.

*Croce:* In our experience with somatic cell hybrids between lymphoblastoid and Burkitt's lymphoma cells, it is clear that the majority of the hybrids express markers typical of lymphoblastoid cells and it seems as though the lymphoblastoid phenotype in that type of cross is dominant.

*Klein:* If you say that reversion occurs because there is a switch to a more controllable phenotype, that is just another way of saying that you get a differentiation switch.

*Bodmer:* I am interested in the HLA class I expression. The relatively low or slightly abnormal expression of these determinants in Burkitt's lymphomas is, in general, fairly clear but it is quite different from the genetic absence of this expression in Daudi. One might suggest that a part of that is due to high levels of *myc* expression. On the other hand, if I understood your revertant data correctly, the revertant still had high levels of *myc* but it had also up-regulated the expression of A11. That suggests that the high level of expression of *myc*, which many people have shown by transfection experiments, is not the cause of the reduced HLA class I expression in Burkitt's lymphomas.

*Klein:* You are absolutely right. Also, one would expect a high level of *myc* expression to be correlated with an overall down-regulation of HLA class I, which we do not see. We see only a very selective down-regulation of some polymorphic HLA determinants.

*Bodmer:* Some experiments were done by a student in my laboratory, Frances Lucibello, with epithelial cells with fairly high levels of expression of exogenous *myc* put into the cells by transfection. We saw about a 50% reduction of HLA class I expression but nothing dramatic. What we thought after the Daudi data was that possibly many Burkitt's lymphomas have lost the expression of one of their 2m genes. In some early published results (Douglass et al 1980), there was often some abnormality in chromosome 15. In principle, one should have checked whether in hybrids one chromosome 15 expresses 2m and the other one doesn't. On the other hand, if it is true for all Burkitt's lymphomas that there is reduced HLA expression, one would expect it to be some secondary consequence of the transformation.

Concerning the renal cell carcinoma, what is intriguing is the very high percentage of deletions in that case. For most of the other cases, including colorectal carcinoma, there have been only a relatively small proportion of deletions reported. It is an interesting question, as to what factors determine, in those cases where one can be reasonably sure there is a recessive effect, whether it is monosomy that starts it off or a deletion or somatic recombination. I feel that selection may play a major role; if you have a large chromosome, monosomy may often be lethal at the cellular level. There is also the possibility of preferential breakpoints that lead to deletions in certain regions; the answer may be a mixture of these factors.

*Klein:* We had two cases of monosomy, but it is rare compared to deletions. There are some preferential sites of deletion, as well, as determined by cytogenetic analysis, the so-called highly fragile sites.

*Bodmer:* I am slightly sceptical of the exact coincidence of those with where the breakpoints occur.

*Nossal:* Concerning the analogy between retinoblastoma and renal cell carcinoma in terms of the visible deletions in the germline, you didn't comment on one significant difference between your age of incidence curves for renal cell carcinoma and those presented this morning by Dr Knudson, that is the very long latent period with an apparent total absence of anything happening until age 40. Why should there be a 40 year latent period before onset of malignancy?

*Klein:* We are inclined to see this in relation to the low rate of cell division in the kidney that may not allow the second change to happen with a reasonable probability before middle age.

*Nossal:* But shouldn't that alter the shape of the curve? Your curves seem to be flat and then suddenly 'take off'.

*Knudson:* I agree, I think it's very difficult. We are looking at some rats with renal carcinoma. We induce the second event with radiation at age three weeks. Microscopic tumours appear at 6–10 months and clinical tumours usually after 18 months. We can't tell whether they occur earlier because if there are only two or four cells, we wouldn't be able to detect them; we must see about 16 cells before we know it is a tumour. The dose–response curve for microscopic tumours in the range 0–900 rads is linear, suggesting one somatic event. So it might be that the event occurs early, but the tumour does not occur until later. I don't know whether that can account for a latency of 40 years in humans. We found the same kind of curve for pheochromocytoma, almost superimposable on what George (Klein) has found, except that the interval is 10-15 years.

*Bodmer:* This is a general problem. It occurs in polyposis, where there is a fairly narrow age of onset distribution, in the teens. One could ask why don't things happen earlier. I don't think that in the colon it has anything to do with rates of cell division, nor do I think one can argue it is waiting for another event. There could easily be environmental factors that are partly responsible. There are also polyposis families with quite different ages of onset of polyposis and it is not yet clear whether the same gene is involved.

*Knudson:* Koji Utsunomiya in Japan has studied the polyposis gene there. Japan has a very high incidence of stomach cancer, and there is a significant incidence of polyps in the stomach in Japan in polyposis patients. We don't see that in the United States and Europe in patients with polyposis.

*Bodmer:* We do see polyps in the upper gastrointestinal tract and one would assume that the difference between that and the colon is not the genetic background but the environment and possible dietary factors.

*Harris:* When you talk about lag periods, you must bear in mind that the second event has to be compatible with cell viability. That second event might occur together with one that is lethal. You might be waiting for a very special situation to get a homozygous recessive that is just right in terms of producing a tumour. I think there's a limit to what you can extract from fitting curves.

Why is the untranslocated *myc* silent, or at least almost silent? I asked Carlo Croce, and he said it is because the cell is not permissive for *myc*, but that doesn't help me much.

*Klein:* Resting cells switch off *myc*. Since the BL cell phenotypically resembles a resting cell, it has already switched to the G0 state, including switching off the non-translocated *myc*, but the translocated *myc* cannot be switched off because it's under the control of the immunoglobulin gene promoter. The immunoglobulin genes continue to be expressed in resting B cells.

*Harris:* A corollary of this would seem to be that once the *myc* gene has been switched off in the resting cell, it remains switched off, even when the cell is restimulated to grow exponentially. The switching off seems, for practical purposes, to be irreversible.

*Bodmer:* But not all dividing cells express *myc*.

*Adams:* Almost all do.

*Klein:* Gerard Evan compared *myc* to the 'dead man's handle' on a British railway train that must always be pushed down when the train is moving. If the driver dies, the handle is released and the train stops.

*Adams:* It seems clear from our results that the affected, i.e. translocated, allele must turn off the other allele through some type of feedback mechanism, as first suggested by Leder et al (1983). We find that at any stage of tumorigenicity where there is constitutive expression of *myc* from an affected allele, for example the Eμ-*myc* transgene, the endogenous alleles are turned off. We have also found that an N-*myc* transgene seems to turn off expression of the endogenous N-*myc* and L-*myc* alleles.

*Harris:* I like that better, because then you achieve a state of functional hemizygosity.

*Varmus:* Was that also true in fibroblasts? Mike Cole has shown that when activated c-*myc* was introduced into fibroblasts the resident alleles were still expressed.

*Adams*: We don't have a lot of data for fibroblasts but the data that we do have show that c-*myc* is turned off there.

*Croce:* That is opposed by evidence from somatic cell crosses between lymphoblastoid and Burkitt's lymphoma cells. If you fuse a Burkitt's lymphoma cell, which expresses only the translocated *myc* allele, with a lymphoblastoid cell that expresses the normal *myc*, the phenotype of the hybrid is lymphoblastoid. If you look at the expression of *myc* by S1 mapping, you find that both *myc* alleles are expressed. This suggests that there is no direct effect of the translocated *myc* on the transcription of the normal *myc*.

*Adams:* The stage-specific expression argument does not appear to fit with our data because at any stage, whether it is a pre-B or B cell or plasma cell, we find that in the presence of an altered allele, the endogenous one is silent.

*Croce:* You can't prove what is responsible for the turning off, you can only prove which one is expressed. If you use a Burkitt's lymphoma cell in which the *myc* is linked to a J region, you see very high expression of the translocated *myc*. That suggests that it is not a direct effect but something else.

*Harris:* When you say expression of both, do you mean expression of all four alleles?

*Croce:* You don't know—you know only about the normal allele compared to the translocated one.

*Harris:* Do you reactivate the allelic inactivated *myc* gene that is not translocated?

*Croce:* That you don't know. You cannot distinguish the normal Burkitt's allele from the normal lymphoblastoid alleles.

*Harris:* You should be able to determine how many of the untranslocated alleles are expressed by gene dosage analysis. You have three copies of the untranslocated allele.

*Nossal:* The phenotype of the Burkitt's cell resembles that of a germinal centre-derived memory B cell. This is curious when we consider that at the time of infection the bone marrow is pumping out huge numbers of virgin B cells, all of which have the EBV receptor. I asked myself, what is one thing that is very special about those germinal centre-derived cells? The one thing is a very high rate of mutation of the immunoglobulin genes. It would appear that these memory type or pre-memory type B cells have a mutation rate of $1 \times 10^{-3}$ bp/generation, which is enormous. Could there be a logical connection between the extraordinary mutation rate and some kind of susceptibility to malignant transformation?

*Knudson:* Does it help to consider that the point of translocation is different in EB-negative and EB-positive cases?

*Klein:* It is not different.

*Croce:* Most of the translocations in endemic cases occur during V-D-J joining, and most of these sporadic translocations might be occuring during switch.

*Klein:* Carlo, would you agree that in all cases it goes to switch except in the high endemic Burkitt's lymphomas? The Burkitt's lymphomas seen in AIDS resemble the sporadic ones in that they rearrange to switch.

*Croce:* The Burkitt's lymphomas in patients with AIDS can go either way, you see V-D-J or switch.

## Reference

Douglass EC, McGrath IT, Lee EC, Whang-Peng J 1980 Serial cytogenetic studies of nonendemic Burkitt's lymphoma cell lines. J Natl Cancer Inst 65:891–895

# Lymphoid neoplasia and the control of haemopoietic differentiation

J.M. Adams, A.W. Harris, W.Y. Langdon, S.P. Klinken, K. Kongsuwan, W.S. Alexander, I. Hariharan, D. Vaux, H. Rosenbaum, M. Crawford, E. Webb, H. Abud and S. Cory

*The Walter and Eliza Hall Institute of Medical Research, Post Office Royal Melbourne Hospital, Victoria 3050, Australia*

*Abstract.* Our broad aims are to delineate oncogenic events in lymphoid neoplasia and to search for genes that control haemopoietic differentiation. To explore lymphoid neoplasia, we have constructed transgenic mice bearing different oncogenes coupled to the immunoglobulin heavy chain enhancer (Eμ), to force expression within lymphocytes. The prototype Eμ-*myc* mice are highly prone to lymphomagenesis, generating pre-B and B cell lymphomas. In their preneoplastic phase, Eμ-*myc* expression perturbs B cell development, accelerating the accumulation of pre-B cells. Lymphomagenesis requires additional oncogenic events, such as *ras* activation, and can be reconstructed *in vitro*. Transgenic mice bearing the N-*myc*, N-*ras*, v-*abl* and *bcr*-v-*abl* oncogenes are also prone to tumours. A striking demonstration that oncogenes can perturb lineage commitment has emerged. Introduction of the v-*raf* gene into cloned Eμ-*myc* transgenic B cells frequently led to a switch in haemopoietic lineage: the cells became macrophages. Two clues to this remarkable metamorphosis are that the macrophage lines produce a myeloid growth factor and most bear marked karyotypic alterations, perhaps indicating that the balance between a few critical lineage control genes has been disturbed. To explore the hypothesis that genes encoding the DNA-binding homeo box domain participate in haemopoiesis, cDNA libraries from haemopoietic sources were screened, and several distinct homeo box cDNAs were isolated. They revealed a complex pattern of expression among haemopoietic cell lines. These genes are attractive candidates for regulators of haemopoietic differentiation.

*1989 Genetic analysis of tumour suppression. Wiley, Chichester (Ciba Foundation Symposium 142) p 54–70*

Although this symposium has the recessive oncogenes as its central theme, this paper considers the contrapuntal theme of the dominant oncogenes. We are attempting to delineate the genetic events that trigger lymphoid neoplasia. To reveal the biological consequences for lymphocytes of the enforced expression of *myc* and other oncogenes, we have constructed transgenic mice in which expression of the introduced gene is targeted to lymphocytes and certain other haemopoietic cells. These mice are providing new insights regarding the influence of oncogenes on differentiation, on oncogene cooperation and on steps in lymphomagenesis.

We have recently also begun to pose questions regarding the nature of the genes that control differentiation processes such as lineage commitment. Some proto-oncogenes may play a role in these, because we have recently found that oncogenes can sometimes annul the commitment of a cell to a particular lineage. To investigate the unknown class of genes that programme cellular commitment, we are exploring the hypothesis that some of these genes are related to those thought to govern embryonic development. It is conceivable that this class of programming or managerial genes will include some of those that can act as recessive oncogenes.

## Transgenic approach to the study of neoplasia

### *Altered myc regulation is tumorigenic*

The hallmarks of murine plasmacytoma and human Burkitt's lymphoma are characteristic translocations that involve chromosomes bearing immunoglobulin (Ig) loci, usually the heavy chain locus (see Cory 1986 for a review). Since their consistent presence in these Ig-synthesizing neoplastic cells argued for a role in tumorigenesis, we attempted to use an Ig gene as a tag to identify the chromosome junction. Using that approach, several groups, including ours (Adams et al 1983), were able to clone the fusion region and to show that the translocation had coupled the IgH locus to the cellular *myc* oncogene. We found that, due to fusion with the highly active Ig locus, the translocated c-*myc* gene was expressed constitutively, while the unaffected allele was silent in every tumour (Adams et al 1983). Thus the major consequence of the translocation is its disruption of normal regulation of *myc* expression. This clear example of activation of a cellular oncogene has provided a paradigm for many of the other karyotypic alterations in tumours (Cory 1986).

Despite the nearly invariant presence of the *myc* translocations, the evidence that they were tumorigenic remained circumstantial. Direct evidence required introducing the gene into the relevant normal cells, preferably within the living organism. We elected to construct transgenic mice (reviewed by Palmiter & Brinster 1986). Although an introduced transgene resides within every cell of the transgenic mouse, its expression can be restricted to certain tissues by the attached regulatory regions. We constructed mice bearing *myc* linked to several regulatory sequences, including those of Ig genes (Adams et al 1985). A transgene derived from a plasmacytoma having the IgH enhancer (Eμ) coupled to c-*myc* proved to be remarkably potent; 13 of 15 primary transgenic mice developed lymphoma within a few months of birth. Moreover, this high predisposition is inherited faithfully. Hence perturbed regulation of this normal cellular gene renders it a potent tumorigenic agent. The specificity for lymphocytes clearly derives from the immunoglobulin enhancer, because expression is confined to lymphoid tissues (Alexander et al 1987) and *myc* linked to other regulatory regions promotes different forms of neoplasia (Stewart et al 1984, Adams et al 1985).

To validate Eμ-*myc* mice as an important new animal model for lymphomagenesis, the incidence and pathology of their disease was evaluated in detail (Harris et al 1988). The mice succumb to disseminated pre-B or B cell lymphoma with an associated leukaemia. However, the disease arises at random within transgenic mice of the same lineage, and nearly all the resulting tumours are monoclonal, indicating that a rare cell has become malignant (see below). The kinetics of tumour onset suggested that a tumorigenic event occurs on average every $10^{10}$ divisions of B lineage cells. The high turnover of early B lymphoid cells may provide a fertile ground for genetic accidents that complement the effects of c-*myc*.

The incidence curves for tumour onset in Eμ-*myc* mice were biphasic (Harris et al 1988). This might reflect genetic heterogeneity, because the mice were an F2 pool derived from the C57BL/6 and SJL strains. Backcrossing to those strains has established that SJL mice bear an allele that enhances susceptibility to the transgene influence (A. Harris, unpublished results). In fact, more than one locus may affect susceptibility to lymphomagenesis.

Conventional BALB/c and NZB mice typically succumb to plasmacytomas six months or more after injection with the mineral oil pristane, and nearly all these tumours bear a *myc* translocation. Since no plasmacytomas had arisen spontaneously in Eμ-*myc* mice, we tested the effect of pristane and found that a substantial proportion did develop plasmacytomas (A. Harris, unpublished results). Hence, *myc* promotes malignancy at all stages of B cell development.

### *Stages in lymphomagenesis*

As stressed in recent reviews (Hanahan 1986, Cory & Adams 1988, Adams 1988), transgenic mice bearing oncogenes provide unique access to the preneoplastic state. The Eμ-*myc* mice, for example, have allowed us to explore how dysregulated *myc* expression perturbs lymphoid development. The great majority (if not all) of the B lineage cells are cycling (Langdon et al 1986). Thus constitutive c-*myc* expression may prevent cells from entering the G0 state. The c-*myc* gene may also regulate cellular differentiation, since the number of cells at early stages of B cell development is elevated about fivefold, while that of Ig-bearing cells is slightly diminished (Langdon et al 1986). Moreover, although the mice remain immunocompetent, fewer Ig-bearing cells respond immunologically, perhaps because the cells are retarded in their development (Vaux et al 1987). These results suggested a model in which increased levels of c-*myc* promote self-renewal at the expense of maturation, whereas reduced levels permit differentiation (Langdon et al 1986).

Although pre-B cells in Eμ-*myc* mice begin to increase above normal numbers even before birth, this initial expansion is benign (Langdon et al 1986). It reaches a plateau when pre-B cells comprise about two-thirds of the marrow — presumably cell death then balances division. The pre-B cells do not appear in notable numbers outside the marrow and spleen, and, by transplantation tests, are not malignant. The conversion to overt, transplantable tumours appears to be associated with ability to invade the lymph nodes and/or thymus, the typical sites of tumour growth (Harris et al 1988).

The behaviour of the pre-neoplastic cells *in vitro* confirms that they are not transformed. Since *myc* is thought to transduce growth factor signals, it is noteworthy that the Eμ-*myc* B lineage cells still require growth factors (Vaux et al 1987, Langdon et al 1988), as also observed for myeloid cells infected with a *myc*-carrying retrovirus (Cory et al 1987). When cultured, the transgenic pre-B cells exhibited much higher cell cycle activity than normal pre-B cells but initially grew to only about twice the normal cell density (Langdon et al 1988). After some weeks, however, they grew to a far higher density, and one culture eventually became fully autonomous and tumorigenic. Thus stages of *myc*-driven lymphomagenesis can be delineated in culture.

### *Oncogene cooperativity*

Considerable evidence favours the view that oncogenesis requires more than one oncogene (reviewed by Weinberg 1985). We find that Eμ-*myc* lymphomas often carry activated oncogenes that can transform NIH-3T3 fibroblasts. At least one tumour contains a mutated N-*ras* gene and another displays an altered K-*ras* gene (W. Alexander & S. Cory, unpublished results). To explore what types of oncogene can cooperate with c-*myc* in lymphomagenesis, we infected neonatal transgenic mice with retroviruses bearing v-H-*ras* or v-*raf*. With either virus, lymphomagenesis was greatly accelerated (W. Langdon, unpublished results).

Particularly dramatic synergy was obtained when Eμ-*myc* mice were bred to mice of the recently constructed Eμ-N-*ras* strain (see below). Indeed the ‘double transgenic’ progeny developed lymphomas within a few weeks of birth. Many of these tumours represent an interesting early stage of B stage development, before any IgH gene has rearranged. We surmise that essentially the first cells exposed to the concerted action of N-*ras* and c-*myc* become fully autonomous. Marked synergy between *myc* and H-*ras* transgenes has also been observed in mammary carcinogenesis (Sinn et al 1987).

The malignant conversion of lymphocytes can now be reproduced in culture (Alexander et al 1989). Both a v-H-*ras* and a v-*raf* virus rapidly engendered a polyclonal proliferation of Eμ-*myc* pre-B cells (Fig. 1), and a substantial proportion of these cells were fully autonomous and grew as tumours in mice. Thus an activated *ras* or *raf* gene can act in concert with

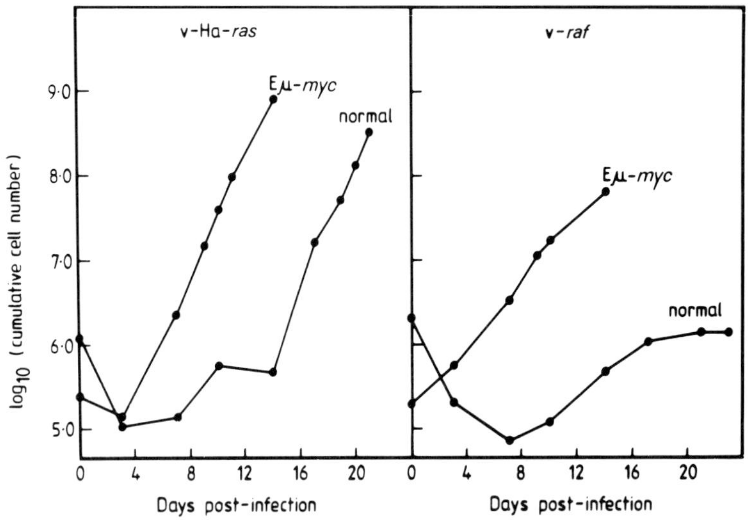


FIG. 1. Cooperation of v-H-*ras* and v-*raf* with c-*myc* in promoting proliferation of B lineage cells. Bone marrow cells from Eμ-*myc* mice and their normal littermates were infected with helper virus-free versions of Harvey murine sarcoma virus, which bears v-H-*ras*, and 3611-murine sarcoma virus (v-*raf*). The infected transgenic cells proliferated indefinitely and were autonomous.

constitutive *myc* expression to fully transform pre-B cells. Surprisingly no such synergy was found with v-*abl* (Dyall-Smith & Cory 1988, Alexander et al 1989), despite its well known ability to transform pre-B cells.

How might oncogenes cooperate in transformation? An appealing theory is that, in order to proliferate, the cell requires signals from two distinct growth factors. Constitutive c-*myc* expression may supplant the need for one of these factors, while an activated *ras* or *raf* gene may replace that for another.

**New transgenic lines**

The insights emerging from the Eμ-*myc* mice have encouraged us to construct transgenic mice bearing other oncogenes subjugated to the IgH enhancer. To facilitate transgene construction, we developed a vector in which the enhancer (Eμ) is coupled to either the SV40 early region promoter or the promoter of an Ig variable region gene, plus the splice signals and polyadenylation sequence from the SV40 T antigen gene. Using these vectors, we have generated mice bearing the N-*myc*, N-*ras*, v-*abl* and *bcr*-v-*abl* genes (Table 1).

**TABLE 1 New transgenic lines**

| *Transgene*[a] | *Tumour type* |
|---|---|
| EμSV N-*myc* | Pre-B lymphomas, B lymphomas |
| EμSV N-*ras*[b] | T lymphomas, macrophage tumours |
| EμSV v-*abl* | T lymphomas, plasmacytomas |
| Eμ$V_H$ *bcr*-v-*abl* | T lymphomas, pre-B lymphomas |

[a] SV indicates the SV40 early region promoter, and $V_H$ the heavy chain variable region promoter.
[b] The *ras* gene is mutated at codon 12.

### *N-myc transgenics and the control of myc expression*

While the structural similarity of N-*myc* and c-*myc* argues for a similar biological role, no alterations of N-*myc* have been observed in any haemopoietic tumour, although the gene has been amplified in many neuroblastomas and certain lung carcinomas. Nevertheless, we find that the effects of N-*myc* upon lymphocytes are similar to those of c-*myc*. The Eμ-N-*myc* mice succumb to either pre-B or B-cell lymphomas (Table 1) and their bone marrow populations are perturbed, albeit less dramatically than in Eμ-*myc* mice. These results argue that N-*myc* and c-*myc* fulfil almost identical biochemical functions.

The c-*myc* gene is thought to be subject to negative feedback regulation, because the normal *myc* alleles are silent in the presence of constitutive *myc* expression from an altered allele (Leder et al 1983, Bernard et al 1983, Adams et al 1985, Alexander et al 1987). It is postulated that an increased level of the *myc* polypeptide leads to a shut down of *myc* transcription. The *myc* transgene, or a translocated *myc* allele, presumably escapes feedback, either because it lacks the responsive DNA sequences or because the negative signal is overridden by the potent immunoglobulin regulatory region.

We find that N-*myc* is regulated similarly. Although N-*myc* is normally expressed in pre-B lymphocytes (Zimmerman et al 1986), the normal N-*myc* alleles are silent in the Eμ-N-*myc* lymphomas. Moreover, as Alt et al (1986) proposed, the regulation apparently extends between N-*myc* and c-*myc*: the c-*myc* gene is completely silent in these tumours and the converse holds for Eμ-*myc* pre-B lymphomas. Thus N-*myc* and c-*myc* appear to be subject to cross-regulation, as well as autoregulation. Moreover, since the Eμ-N-*myc* pre-B lymphoma cells can proliferate without any detectable c-*myc* mRNA (and *vice versa*), the two gene products must be interchangeable in their biochemical function(s), at least within pre-B cells.

### *N-ras induces two types of tumours*

N-*ras* is frequently mutated in certain haemopoietic neoplasms, particularly in acute myeloid leukaemia of humans and some T lymphomas of mice.

Interestingly, transgenic mice bearing an N-*ras* gene mutated at codon 12 develop two types of tumours: thymic T lymphomas and a histiocytic tumour, apparently of macrophage origin, that greatly enlarges the spleen and liver. To our surprise, no tumours of B cell origin have appeared. Since breeding to Eμ-*myc* mice greatly accelerated development of B lineage tumours (see above), the Eμ-N-*ras* transgene must be expressed within B lineage cells. Hence N-*ras* alone must be relatively innocuous within B cells.

### *v-abl and bcr-v-abl*

Since the tumour type most frequently generated by Abelson murine leukaemia virus is a pre-B lymphoma, we were very surprised to find that Eμ-v-*abl* mice yield T lymphomas and plasmacytomas but no pre-B or B lymphomas. An intriguing explanation could be that the transforming activity of v-*abl* is confined to certain stages of B cell development, one being the plasma cell and the other an extremely early stage before the pre-B cell. At that early stage, the IgH enhancer might not yet be active and hence the v-*abl* transgene would be silent.

In the chromosome translocation that generates the small 'Philadelphia' chromosome, the cellular *abl* gene becomes fused to the previously unknown *bcr* (breakpoint cluster region) gene (reviewed by Cory 1986). The resulting hybrid *bcr-abl* gene is of particular interest because this translocation is strongly associated with chronic myeloid leukaemia and some cases of acute lymphoid leukaemia. A full length *bcr* cDNA (Hariharan & Adams 1987) allowed us to generate a synthetic *bcr*-v-*abl* gene that resembles the *bcr*-c-*abl* gene. The Eμ-*bcr*-v-*abl* mice have succumbed to lymphomas, perhaps mimicking the acute lymphoblast proliferation (lymphoid blast crisis) that sometimes terminates the chronic phase of myeloid leukaemia. An important clue to the function of the *bcr-abl* gene is that it can render IL-3-dependent cell lines autonomous and tumorigenic by a mechanism involving production of trace amounts of IL-3 (Hariharan et al 1989). Hence autocrine factor production may be central to the development of chronic myeloid leukaemia.

## Genetic control of haemopoietic differentiation

Given the complexity of lineage commitment and maturation within the lympho-haemopoietic system, we surmise that these processes must be controlled genetically by a small number of master genes. The only lineage control gene so far described in a vertebrate is one that governs muscle cell development (Davis et al 1987).

### *Induced lineage switch*

Oncogenes may affect cellular commitment, because we have recently found

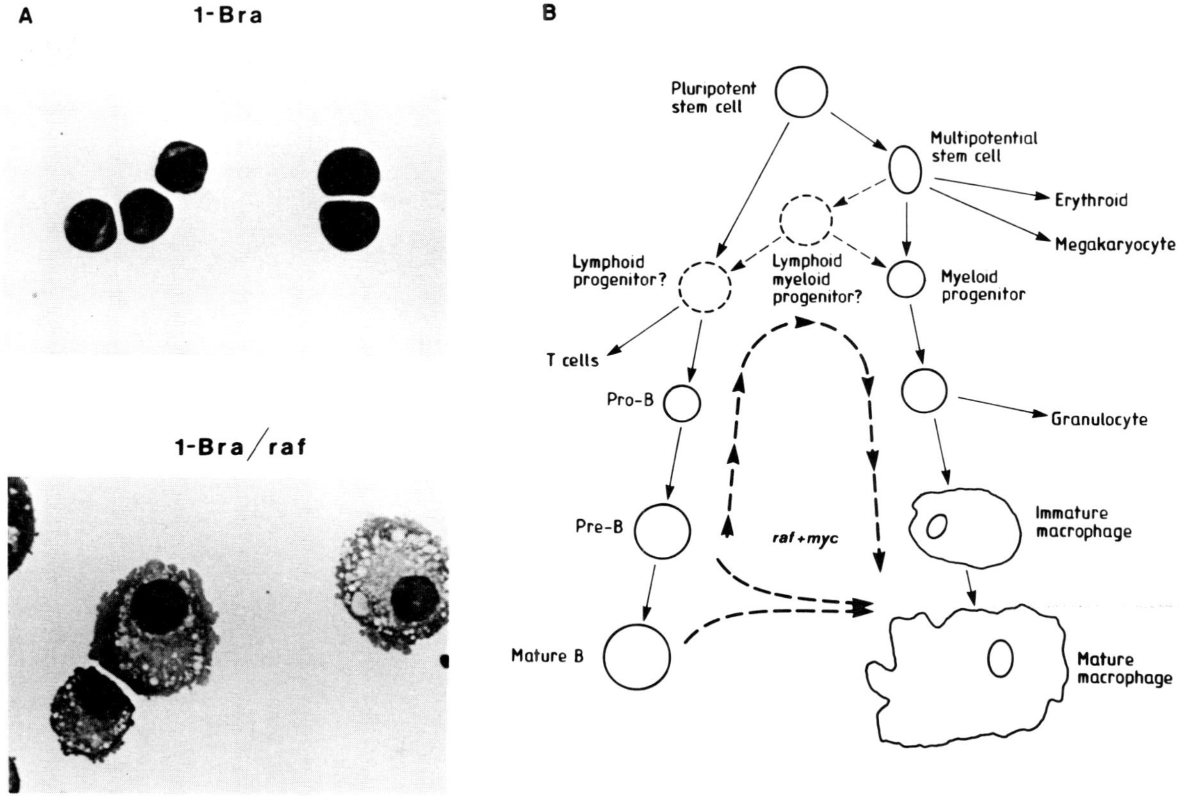


FIG. 2. Haemopoietic lineage switch. A) The change from a lymphoid morphology in pre-B line 1-Bra to the macrophage 1-Bra/raf. B) The relationship of the relevant haemopoietic lineages (from Klinken et al 1988).

that an oncogene can provoke a striking lineage conversion. Introduction of v-*raf* into cloned Eμ-*myc* pre-B or B lymphoma cell lines often generated cells with all the features of macrophages (Klinken et al 1988). Their altered morphology, for example, is illustrated in Fig. 2A. The lymphoid phenotype was lost, but the cells retained their idiosyncratic Ig gene rearrangements. Thus, as indicated in Fig. 2B, cells well along the B-cell differentiation pathway have changed into cells of the myeloid lineage, either by a direct switch or via retrodifferentiation to a common precursor.

Two clues to this remarkable metamorphosis are that a myeloid growth factor is produced by all the derivative lines and that most of these lines display gross karyotypic alterations. Since some lymphoid lines expressing both *myc* and v-*raf* do not switch lineage, the reprogramming must be at least partly indirect. We speculate that the *myc*/*raf* combination induces some karyotypic instability and that this sometimes alters the balance between a few master genes that programme the cell to differentiate to a particular cell lineage.

### *Expression of multiple homeo box genes in haemopoietic cells*

To explore the possibility that haemopoietic differentiation is regulated by genes akin to those thought to govern embryogenesis, we have tested whether haemopoietic cells express genes bearing the homeo box, the conserved DNA-binding domain first identified in morphology-determining genes of *Drosophila* (reviewed by Gehring 1987). Fig. 3 shows that multiple transcripts of homeo box genes can be detected in haemopoietic cell lines, and we have obtained several homeo box cDNA clones from haemopoietic tissue of mice and a human leukaemia cell line (Kongsuwan et al 1988). These include different known genes as well as previously undescribed ones. They display diverse patterns of expression, some of which alter when differentiation is induced. These results raise the possibility that some homeo box genes are controlling elements in haemopoiesis. They may, for example, control the large sets of structural genes that define cell phenotype and/or proliferation status.

If homeo box genes govern the terminal differentiation of cells, perturbation of their expression could block programmed cell death and thereby contribute to neoplasia. It is interesting to note that a myeloid leukaemia cell line has an altered homeo box gene (Kongsuwan et al 1988). Hence certain homeo box genes may prove to represent a new class of oncogene, which could be of either the dominant or recessive type.

### *Acknowledgements*

This research was supported by the National Health and Medical Research Council (Canberra) and by the U.S. National Cancer Institute (grants CA12421 and CA43540).

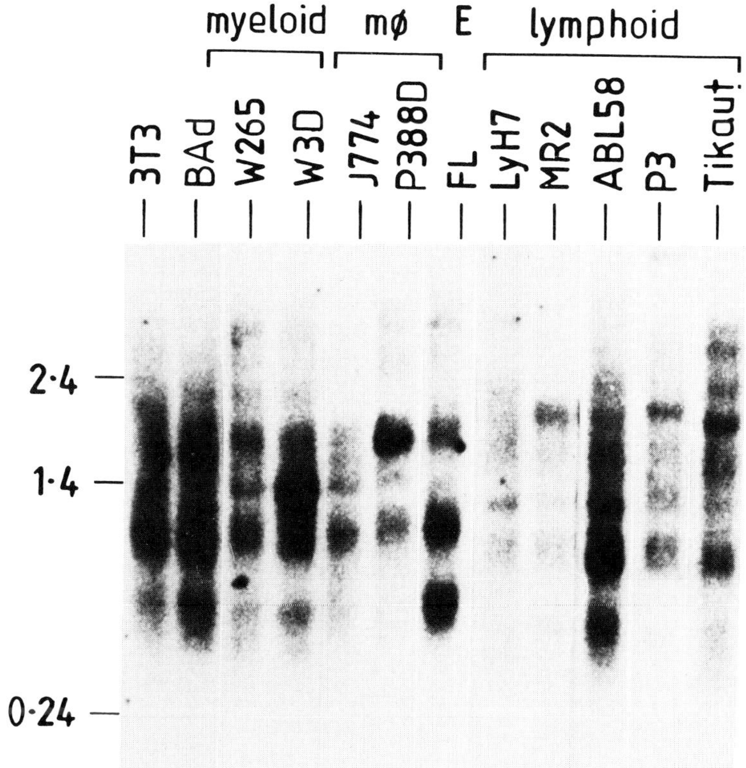


FIG. 3. Expression of multiple homeo box genes in haemopoietic cell lines. On this Northern blot of poly A$^+$ mRNA from cells lines representative of early myeloid, macrophage (mΦ), erythroid (E) and T and B lymphoid cells, homeo box transcripts were revealed by hybridization at low stringency with oligonucleotide probes from the most conserved region of the homeo box (Kongsuwan et al 1988).

## References

Adams JM 1988 Consequences of constitutional activation of oncogenes in transgenic mice. In: Klein G (ed) Cellular Oncogenes, Marcel Dekker, New York

Adams JM, Gerondakis S, Webb E, Corcoran LM, Cory S 1983 Cellular *myc* oncogene is altered by chromosome translocation to an immunoglobulin locus in murine plasmacytomas and rearranged similarly in Burkitt lymphomas of man. Proc Natl Acad Sci USA 80:1982–1986

Adams JM, Harris AW, Pinkert CA et al 1985 The c-*myc* oncogene driven by immunoglobulin enhancers induces lymphoid malignancy in transgenic mice. Nature (Lond) 318:533–538

Alexander WS, Schrader JW, Adams JM 1987 Expression of the c-*myc* oncogene under the control of the immunoglobulin enhancer in Eμ-*myc* transgenic mice. Mol Cell Biol 7:1436–1444

Alexander WS, Adams JM, Cory S 1989 Oncogene cooperation in lymphocyte transformation. Mol Cell Biol, in press

Alt FW, DePinho R, Zimmerman K et al 1986 The human *myc* gene family. Cold Spring Harbor Symposium, Cold Spring Harbor, New York 51:931–941

Bernard O, Cory S, Gerondakis S, Webb E, Adams JM 1983 Sequence of the murine and human cellular *myc* oncogenes and two modes of *myc* transcription resulting from chromosome translocation in B lymphoid tumours. EMBO J 2:2375–2383

Cory S 1986 Activation of cellular oncogenes in haemopoietic cells by chromosome translocation. Adv Cancer Res 47:189–234

Cory S, Adams JM 1988 Transgenic mice and oncogenesis. Annu Rev Immunol 6:25–48

Cory S, Bernard O, Bowtell D, Schrader S, Schrader JS 1987 Murine c-*myc* retroviruses alter the growth requirements of myeloid cell lines. Oncogene Res 1:61–76

Davis RL, Weintraub H, Lasser AB 1987 Expression of a single transfected cDNA converts fibroblasts to myoblasts. Cell 51:987–1000

Dyall-Smith D, Cory S 1988 Transformation of bone marrow cells from Eμ-*myc* transgenic mice by Abelson murine leukemia virus and Harvey murine sarcoma virus. Oncogene Res 2:403–409

Gehring WJ 1987 Homeo boxes in the study of development. Science (Wash DC) 236:1245–1252

Hanahan D 1986 Oncogenesis in transgenic mice. In: Kahn P, Graf T (eds) Oncogenes and Growth Control, Springer-Verlag, Heidelberg

Hariharan I, Adams JM 1987 cDNA sequence for human *bcr*, the gene that translocates to the *abl* oncogene in chronic myeloid leukaemia. EMBO J 6:115–119

Hariharan IK, Adams JM, Cory S 1989 *bcr-abl* oncogene renders myeloid cell line factor independent. Oncogene Res, in press

Harris AW, Pinkert C, Crawford M, Langdon W, Brinster R, Adams JM 1988 The Eμ-*myc* transgenic mouse: a model for high incidence spontaneous lymphoma and leukemia of early B cells. J Exp Med 137:353–371

Klinken P, Alexander W, Adams JM 1988 Hemopoietic lineage switch: v-*raf* oncogene converts Eμ-*myc* transgenic B cells into macrophages. Cell 53:857–867

Kongsuwan K, Webb E, Housiaux P, Adams JM 1988 Expression of multiple homeo box genes within diverse haematopoietic lineages. EMBO (Eur Mol Biol Organ) J 7:2131–2138

Langdon WY, Harris AW, Cory S, Adams JM 1986 The c-*myc* oncogene perturbs B lymphocyte development in Eμ-*myc* transgenic mice. Cell 47:11–18

Langdon WY, Harris AW, Cory S 1988 Growth of Eμ-*myc* transgenic B-lymphoid cells *in vitro* and their evolution towards autonomy. Oncogene Res 3:271–279

Leder P, Battey J, Lenoir G et al 1983 Translocation among antibody genes in human cancer. Science (Wash DC) 222:765–771

Palmiter RD, Brinster RL 1986 Germline transformation of mice. Annu Rev Genet 20:465–499

Sinn E, Muller W, Pattengale P et al 1987 Coexpression of MMTV/v-H-*ras* and MMTV/c-*myc* genes in transgenic mice: synergistic action of oncogenes *in vivo*. Cell 49:465–475

Stewart TA, Pattengale PK, Leder P 1984 Spontaneous mammary adenocarcinomas in transgenic mice that carry and express MMTV/*myc* fusion genes. Cell 38:627–637

Vaux D, Adams JM, Pike B 1987 Immunological competence of B cells subjected to constitutive c-*myc* oncogene expression in Eμ-*myc* transgenic mice. J Immunol 139:3854–3860

Weinberg RA 1985 The action of oncogenes in the cytoplasm and the nucleus. Science (Wash DC) 230:770–776

Zimmerman KA, Yancopoulos GD, Collum RG et al 1986 Differential expression of *myc* family genes during murine development. Nature (Lond) 319:780–783

## DISCUSSION

*Stanbridge:* In the *myc* transgenic mouse infected with *ras*, there was synergism and the very rapid appearance of tumours. Are those tumours clonal?

*Adams:* Yes, at least some are. We are not convinced, however, that this means a third event is necessary for the tumorigenic phenotype. Even if a cell is tumorigenic, there may still be effects on growth rate, for example, over and beyond that. If we infect Eμ-*myc* pre-B cells *in vitro* with a *ras*-carrying virus, such that a high proportion of cells have been infected, on continued growth of those cells or their introduction into a mouse, we still get essentially clonal lines and tumours from the polyclonal population. We think this means subtle changes, that augment growth rate, for example, will give certain cells an advantage in a competitive situation.

*Varmus:* I assume you superinfect the transgenic mice using stocks containing helper virus?

*Adams:* No, we use helper-free virus.

*Varmus:* Is it fair to derive anything from the observation that these tumours appear to be clonal? How did you establish clonality?

*Adams:* By immunoglobulin gene rearrangements and by the number of proviral integration sites.

*Varmus:* It is possible that 50 independent cells contributed to the initial growth of the tumour, but by the time you analyse the tumour one predominates as a result of competition. There could be clonal dominance but it doesn't mean that the tumour arose from a single cell.

*Bodmer:* I suspect that's not true. You are talking about differential selection in a cell population. If there were no selective differences between the subclones in an outgrowing population, during the time that one is looking there would not be fixation of one type rather than another. I would support Jerry's (Adams) guess that there is some growth rate advantage on top of everything else.

*Varmus:* Do you know how many cells are infected in these experiments? How many infectious units are delivered to the mouse?

*Adams:* It is of the order of $10^3$ to $10^5$ infectious units, as measured by focus-forming units on 3T3 fibroblasts.

*Bodmer:* It depends on the number of founder cells. If that's of the order of the number of generations, then you have a fair chance of fixation. I assume the number of generations is in the tens, so if you go to fixation there will be two consequences: either there is effectively a small starting population, as in the example of re-populating the bone marrow, or there is selection. You can't tell which is which until you have determined the starting conditions. But if your starting population is nearly $10^5$ cells, then fixation due to random effects is exceedingly unlikely. So either there are other events giving rise to selection, or you have to argue that in practice a very small proportion of the transferred cells are contributing to the growth of the tumours.

*Nossal:* Jerry (Adams) has a very easy way of testing that question; easy in concept, more difficult in execution. He has a three week window during which there are no frank malignancies in the double *ras/myc* transgenics. If he does a limit dilution experiment, taking those cells and putting them into nude mice, sooner or later at the limit dilution tumours will appear in these adoptive hosts. Then he could determine whether clonality arises because of a further event or whether in the original animal one cell got past the post first.

*Adams:* We plan to try something similar to that, but probably *in vitro* first. We are going to put the *myc ras* transgenic pre-B cells into an agar cloning system in which the normal cells are not able to grow as colonies. Then we can find out whether a high proportion of the transgenic cells will generate autonomous clones.

I still feel that the question of whether, say, *myc* plus *ras* is sufficient has not been fully answered. Bob, what do you think about the evidence that *myc* plus *ras* apparently does not readily transform human cells? Has that sort of experiment been done well enough? If that conclusion is justified, I find it disturbing because *myc* and *ras* are likely to have extremely general functions in the cell, irrespective of species.

*Weinberg:* You pose two questions: the first concerns whether *ras* and *myc* are sufficient to transform rodent fibroblasts into which they have been introduced by transfection or infection. There I have no strong opinions. The work of Carl Barrett suggested that in addition to the combined actions of a *ras* and a *myc* oncogene, a subsequent shift in chromosomal constitution may be required. Therefore *ras* and *myc* may be necessary but not sufficient for full transformation of these cells.

The second question concerns human cells, which are a great mystery to me. None of us believe that human cells are in any way fundamentally different from rodent cells. However, there is a large volume of evidence which suggests that they are much harder to transform. I don't know why that is; I suspect that it has nothing to do with the *ras* or *myc* genes *per se*. That doesn't reduce the importance of *ras* or *myc*; instead, it means that there is an additional level of complication in human cells. Perhaps there is a physiological barrier to transformation in human cells that is quantitatively but not qualitatively different from a comparable entity in rodent cells.

*Bodmer:* I think there is a simple answer if you are talking about primary cells. The repair proficiency is greater in primary human cells and there is probably therefore a lower mutation rate. That argues in favour of the relevance of further events.

*Weinberg:* I don't accept that. If you put *ras* and *myc* into rodent cells *in vitro*, you get tumorigenic cell clones growing out almost immediately, independent of rare tertiary mutations that might be occurring.

*Bodmer:* How do you know about the background of the mutations that are already there in what grows up? Even primary cells will have a certain rate at which mutations happen.

*Weinberg:* I think you may be grasping for straws!

*Bodmer:* I think it is a fundamental question that's related to the age incidence of cancer.

*van der Eb:* Morphologically transformed cells cannot even be obtained with *myc* and *ras* in diploid fibroblasts, I believe.

*Lau:* We transfected human diploid fibroblasts with both c-*myc* and the human mutant c-Ha-*ras* gene EJ, using pSV*gpt* and pSV*neo* vectors (Sager et al 1983). The transfected cells were recovered by selection with either mycophenolic acid (for the *gpt* gene) or G418 (for the *neo* gene). We found that these transfected oncogenes were expressed at high levels, as shown by monoclonal antibody staining of their products. But these cells did not make tumours in nude mice, and they were not even morphologically transformed.

*Harris:* Jerry, these rapidly growing tumours into which you have put a second gene, are they chromosomally the same as the slowly growing tumours that don't have the second gene? If they are chromosomally different, you have a way out your numerical problem, because certain kinds of chromosomal event occur with a much greater frequency than classical mutations. For example, in an aneuploid cell, non-disjunction can occur much more frequently than 1 in $10^6$. You would not necessarily find systematic differences between the two kinds of tumour, but are the cells that receive the second gene grossly different in their chromosome constituents from those that don't?

*Adams:* That hasn't been looked at in detail yet. The *ras*-infected Eμ-*myc* pre-B cells exhibit no gross changes, the cells are diploid with no obvious marker chromosomes.

*Weinberg:* You made a remark dissociating *myc* action from immortalization. For me immortalization is much more an *in vitro* phenotype that's dependent on culture conditions, such as serum factors. Why do you imply that there are no immortalizing effects here?

*Adams:* We think that *myc* does not immortalize lymphocytes, because we don't see the rapid emergence of cell lines from cultures of Eμ-*myc* pre-B cells maintained on stromal layers (Langdon et al 1988). On long term maintenance in culture, say 16–20 weeks, we have observed the emergence of a cell line in only a single instance. Whether such a rare event should be termed an immortalizing function seems problematical.

*Marks:* In our studies with murine erythroleukaemia (MEL) cells there seems to be a correlation between induction of terminal differentiation by hexamethylene *bis*acetamide and suppression of *myb* expression. In these cells, continued expression of *myb* is associatiated with proliferation; suppression of the *myb* gene is associated with the expression of differentiated characteristics and cessation of proliferation (Marks et al 1987). Have you looked at expression of *myb* in the B cells?

*Adams:* We have to some extent. As others have reported, *myb* expression seems to depend on the stage of development—it is reasonably high in the pre-B cells, but falls to an extremely low level as they mature. I agree that *myb* may play an important role in differentiation. I agree also that in some non-lymphoid lineages it's not obvious whether *myc* has much impact on differentiation.

*Varmus:* Dr Marks, have you tried forcing expression of *myb* with a constitutive promoter in your erythroleukaemia cells?

*Marks:* Yes, transfection of MEL cells with c-*myb* can cause expression of high levels of *myb* and make the cells resistant to inducer (Clark et al 1988).

*Bodmer:* On a different issue, for a lot of these controlling proteins, for example the DNA-binding proteins, it seems to me that one doesn't have the classic situation of dominance and recessivity due to essentially excessive gene product in a hemizygote. The cell is probably nearly always producing a relatively limiting amount of gene product so that a dosage effect becomes important. In other words, reducing the level of gene product by 50% for some of those controlling proteins may give an effect, as we suggested for the colorectal situation, when there is stochastic variation. That leads to competition for binding of those proteins. This could explain the translocated versus the non-translocated *myc*. If those sorts of gene products are involved in a linear sequence of switches, you may easily switch things by a delicate balance in the level of expression. It also slightly confuses the rather simple dominance/recessive interpretations of phenotype. One has to think in terms of the possibility of low levels of gene product, where small variations due to gene dosage, which is essentially what happens in the hybrid, can be quite significant.

*Klein:* Richard Mulligan and Dr Bernstein found that repopulation with normal bone marrow leads to takeover by a single clone. By analysis of retroviral insertion site, they found that a different clone grows out from the same polyclonal bone marrow culture in different mice. You don't need to postulate any specific event to explain clonality, except that one clone gets ahead of the others.

In many B cell neoplasias, like Waldenström's disease or human plasmacytomas, there is a strong inhibition of normal B cell function. This suggests a negative feedback that should be taken into consideration. Apparently, when one B cell clone gets ahead, others are inhibited.

*Bodmer:* George, don't your data imply that the population of bone marrows would be a cohort? It seems that there would be a problem of adequate repopulation of the bone marrow if there was too much of a tendency for a small number of clones to grow out.

*Adams:* I am not sure how relevant bone marrow repopulation is. That probably reflects the normal control of haemopoiesis—a homeostatic mechanism may operate to ensure that only a very small number of stem cell clones are used in the animal at any one time.

*Klein:* But only one is used in these tumours.

*Varmus:* The setting is very different in repopulation experiments, because the animals have been irradiated. You are demanding that bone marrow cells be derived from the few infected stem cells in the population that's being returned to the animal.

*Noda:* Has anybody looked at the chromosomal constitution of those tumours? Oshimura et al (1985) have shown with fibroblasts that there is loss of specific chromosome(s).

*Adams:* The lymphomas that arise in our Eμ-*myc* mice have been examined by Dr Francis Wiener in the Karolinska Institute. Certain karyotypic abnormalities have been found, but there is no consistent pattern.

*Klein:* There are no *myc* translocations and the anomalies are essentially random.

*Harris:* To me, the important thing in experimental systems like Harold Varmus' or Jerry Adams' is not really whether the tumours are clonal, but what degree of selection is required to produce them. If you put a gene into $10^6$ cells and only one cell actually generates the tumour, the specificity of the effect is not much better than can be achieved with X-irradiation. If 70% of the cells give rise to the tumour, then you are obviously doing something highly specific. You do not have to have a formal demonstration that the tumour actually arose from one particular cell. There is a much more primitive test that is useful. You can simply compare the karyotypes of the cells that constitute the tumour with those of the cells that you started with. If they are grossly different, that bears very heavily on the interpretation of the experiments. It is essential to know whether you are producing a tumour in a direct sort of way or whether you are simply producing a population of cells in which an odd one can generate a tumour. It seems to me that much more important than the formal question of clonality is the degree of selection in the experiment.

*Noble:* I would like to describe some experiments done by Hartmut Land and Timothy Thompson, which relate to these issues. The experiments start with dissatisfaction with transgenic animals as models for tumorigenesis, because in transgenics you are activating an oncogene in every cell within a particular lineage. Their approach is to take embryonic prostate, dissociate it in culture and infect it with a retrovirus. The retrovirus contains either *myc*, *ras* or a *ras*/*myc* combination. The virus can be plus or minus β-galactosidase for visualization of infected cells.

If you put in *ras* by itself, then reconstitute the tissue by implanting under the kidney capsule, at a very low rate of infectivity you can see cells expressing β-galactosidase, but the tissue is normal. At a slightly higher rate of infectivity, you see a few more β-galactosidase positive cells and there is mesenchymal dysplasia. Very importantly for consideration of pathology of tumours, tissue can be dysplastic with as few as 5% of the cells expressing β-galactosidase, therefore presumably as few as 5% of the cells expressing a *ras* oncogene.

With the same virus, infecting the same tissue in exactly the same way, but using *myc*, there is no mesenchymal dysplasia but an epithelial hyperplasia. The implication is that the pre-neoplastic phenotype is the result of a single oncogenic hit, at least in this sort of tissue.

Putting in *ras/myc* together, obviously you get tumours but the interesting question is, is it a two-hit process or a three-hit process? That has been addressed by looking over a range of infectivities at which tumours occur. At a rate of infection about 10- fold higher than the minimum needed to see any tumours, within the reconstituted prostate the tumours are predominantly clonal. So even though there may have been 20 or 200 cells successfully infected with the *ras/myc* virus, only one of those cells grows out to predominate. These results argue that there is a third event.

## References

Clarke MF, Kukowska-Latallo JF, Westin E, Smith M, Prochownik EV 1988 Constitutive expression of a c-*myb* cDNA blocks Friend murine erythroleukemia cell differentiation. Mol Cell Biol 8:884–892

Langdon WY, Harris AW, Cory S 1988 Growth of Eμ-*myc* transgenic B-lymphoid cells *in vitro* and their evolution towards autonomy. Oncogene Res, in press

Marks PA, Sheffery M, Rifkind RA 1987 Induction of transformed cells to terminal differentiation and the modulation of gene expression. Cancer Res 47:659–666

Oshimura M, Gilmer TM, Barrett JC 1985 Nonrandom loss of chromosome 15 in Syrian hamster tumours induced by v-Ha-*ras* plus v-*myc* oncogenes. Nature (Lond) 316:636–639

Sager R, Tanaka K, Lau CC, Ebina Y, Anisowicz A 1983 Resistance of human cells to tumorigenesis induced by cloned transforming genes. Proc Natl Acad Sci USA 80:7601–7605

# General discussion I

## Suppression of tumorigenicity by polar compounds that induce terminal differentiation of transformed cells[1]

*Marks:* We have been studying the ability of low molecular weight (generally Mr~400) polar compounds to induce murine erythroleukaemia (MEL) cells to express the differentiated phenotype, including cessation of proliferation (Marks et al 1987, Marks & Rifkind 1978). These polar compounds can induce a large variety of transformed cells to differentiate and lose the capacity to proliferate (Table 1). Friend and colleagues (Friend et al 1971) demonstrated that virus-transformed MEL cells in culture with dimethysulphoxide are induced to erythroid differentiation. We investigated the relationship between the structure and activity of inducing compounds (Tanaka et al 1975) and found the polar group to be critical for inducer activity. We identified a new group of potent inducers, the polymethylene bisacetamides, of which hexamethylene bisacetamide (HMBA) is among the most potent (Fig. 1) (Tanaka et al 1975, Reuben et al 1976, 1978).

**TABLE 1 (*Marks*) Transformed cell lines which can be induced to differentiate by HMBA or other polar compounds**

| *Cell lines* | *Primary cell cultures* |
|---|---|
| Erythroleukaemia (mouse) | Colon carcinoma (human) |
| Teratocarcinoma (mouse) | Bladder carcinoma (human) |
| Hepatic tumour (mouse) | Acute myeloid leukaemia (human) |
| Neuroblastoma (mouse) | Melanoma (human) |
| Mammary tumour (rat) (human) | |
| LB myeloblast (rat) | |
| Kidney epithelial carcinoma (canine) | |
| HL-60 promyelocytic leukaemia (human) | |
| Melanoma (human) | |
| Colon carcinoma (human) | |
| Bladder carcinoma (human) | |
| Glioblastoma multiforma (human) | |
| T Cell (human) | |
| B Cell (human) | |

For references, see review by Marks et al (1987).

[1] The studies discussed which were performed in our laboratories were partly supported by grants from the National Cancer Institute (CA-31768 and CA-08748).

**POLAR COMPOUNDS INDUCING MELC DIFFERENTIATION**

| | | CONCENTRATION (mM) | CELL DENSITY ($\times 10^5$/ml) | % B+ |
|---|---|---|---|---|
| N-METHYLACETAMIDE | $CH_3$-C(=O)-NH $CH_3$ | 30 | 1.2 | 32 |
| N,N-DIMETHYLACETAMIDE | $CH_3$-C(=O)-N($CH_3$)$CH_3$ | 30 | 1.9 | 60 |
| N,N-DIMETHYLFORMAMIDE | HC(=O)-N($CH_3$)$CH_3$ | 105 | 1.3 | 63 |
| DIMETHYLSULFOXIDE | $CH_3$-S(=O)-$CH_3$ | 280 | 0.8 | 82 |
| HEXAMETHYLENE BISACETAMIDE | $CH_3$-C(=O)-N(H)-$(CH_2)_6$-N(H)-C(=O)-$CH_3$ | 5 | 1.6 | 98 |
| HEXAMETHYLENE BIS-METHYL CARBOXAMIDE | $CH_3$-N(H)-C(=O)-$(CH_2)_6$-C(=O)-N(H)-$CH_3$ | 5 | 1.7 | 95 |

FIG. 1. (*Marks*) Relative activity of several polar compounds as inducers of MEL cell (DS19) terminal differentiation. Cell growth and percentage of benzidine positive cells were determined after five days in culture with compound at the optimal concentration indicated (Tanaka et al 1975, Reuben et al 1978).

HMBA-mediated MEL cell terminal differentiation is a multistep process. There is an early, 'latent' period of approximately 10 to 12 hours during which a number of metabolic changes occur but no commitment to terminal differentiation can be detected (Fig. 2A) (Marks et al 1987). We define commitment as the capacity to express characteristics of the differentiated erythroid phenotype, including loss of proliferative capacity, despite removal of the inducer (Gusella et al 1976, Fibach et al 1977). With continued exposure to HMBA, there is a transition from the latent period to a period when there is recruitment of an increasing proportion of the cell population to terminal erythroid differentiation with loss of proliferative capacity.

*HMBA-induced MEL cell differentiation in vitro: characteristics of latent period*

During the latent period, HMBA causes alterations in membrane fluidity; changes in membrane permeability to a number of ions, including sodium, potassium and possibly calcium; a prompt increase in membrane-associated protein kinase C (PKC) activity; and the appearance in the cytosol of a $Ca^{2+}$- and phospholipid-independent form of PKC, presumably generated by proteolytic cleavage of membrane-bound PKC; and modulation of the expression of

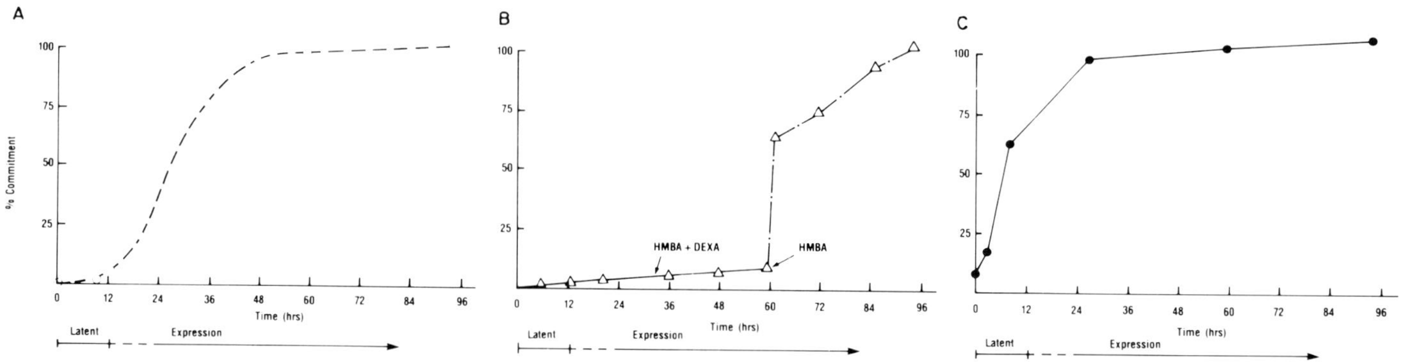


FIG. 2. (*Marks*) (A) Kinetics of HMBA induction of MEL cell (DS19) to terminally differentiate. Commitment was assayed as described in Fibach et al 1977. (B) Dexamethasone (DEXA) blocks the expression of HMBA-induced MEL cell (DS19) commitment to terminal differentiation (Chen et al 1982). (C) Kinetics of HMBA induction of vincristine-resistant MEL cells (V3.17) to terminal differentiation (Melloni et al 1988).

a number of genes, among which are c-*myb*, c-*myc*, c-*fos* and p53 (Marks et al 1987).

The level of c-*myc* RNA decreases markedly within the first hour of culture with HMBA. A decrease in c-*myb* mRNA level occurs within 1–4 hours. c-*fos* mRNA is almost undetectable in uninduced MEL cells; in cells cultured with HMBA it begins to accumulate within about four hours, reaching a maximum by 18 to 24 hours. p53 protein decreases appreciably by eight hours. Dexamethasone (Fig. 2B) and TPA (Fibach et al 1977) each prevent expression of HMBA-induced MEL cell terminal differentiation but do not modify the changes in c-*myc*, c-*myb*, c-*fos* or p53 gene expression observed during the latent period in cells cultured with HMBA alone (Ramsay et al 1986).

We have reported that DS19 cells cultured with dexamethasone and HMBA for up to 60 hours do not express commitment (Chen et al 1982). Removal of the steroid and continued culture with HMBA is associated with recruitment of a substantial proportion of the cells to differentiate without a latent period (compare Fig. 2A with 2B). These observations suggest that the steroid-sensitive step which blocks HMBA-induced commitment occurs late in the latent period.

### *HMBA-induced MEL cell commitment to differentiation*

Continued culture of MEL cells with HMBA beyond the latent period results in morphological and chemical changes similar to normal terminal erythroid differentiation. These include increased expression of a number of genes, including genes for $\alpha^1$ and $\beta^{maj}$ globin, the haem synthetic enzymes and erythroid-specific membrane proteins, as well as suppression of DNA replication and of rRNA synthesis (Marks & Rifkind 1978, Tsiftsoglou & Robinson 1985, Marks et al 1987).

After the decrease in c-*myc* mRNA that occurs during the early part of the latent period, the amount of the message returns to the level in uninduced cells by about 12 hours and remains there for at least the ensuing 36 hours; the elevated level of c-*fos* mRNA remains essentially unchanged. By comparison, by 12 hours the level of c-*myb* mRNA increases towards control levels from its early nadir, then falls again to undetectable levels by 24 to 36 hours, where it remains for the duration of culture with inducer. Inducer-mediated suppression of c-*myb* mRNA may be an important factor in the recruitment of cells to commitment to terminal cell division, as suggested by several lines of evidence: (1) continued suppression of c-*myb*, but not of c-*myc*, is associated with progressive recruitment of MEL cells to terminal cell division; (2) haemin, which induces globin mRNA accumulation but not commitment to terminal cell division, does not suppress transcription of c-*myb* mRNA; (3) dexamethasone inhibits HMBA-induced commitment and inhibits the late (>12 hours) suppression of c-*myb* mRNA, but does not modify the changes in expression of c-*myc*, c-*fos* or p53 genes that occur in response to HMBA; (4) transfection of MEL cells with the c-*myb* gene and expression of the exogenous *myb* DNA is associated with resistance to polar compound-induced differentiation (Clarke et al 1988).

*Suppression by HMBA of MEL cell oncogenicity*

One of the most striking findings in MEL cells induced to differentiate with HMBA is suppression of oncogenicity of these leukaemic cells on inoculation of pre-treated cells into susceptible mice (Marks et al 1978). In our laboratory, inoculation of MEL cells caused leukaemia with death in under four weeks (Fig. 3). Inoculation of up to five times more MEL cells pre-cultured for 100 hours with HMBA almost completely suppressed oncogenicity.

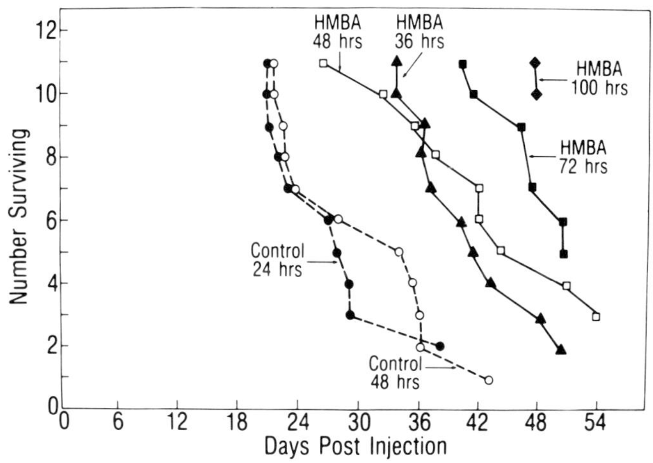


FIG. 3 *(Marks)* Effect of pre-incubation with HMBA on oncogenicity of MEL cells, as measured by survival of mice after inoculation of cells. For the controls, each mouse was inoculated with 2 x $10^5$ MEL cells that had been cultured for five days without HMBA. Animals receiving cells pre-cultured with HMBA were inoculated with 1 x $10^6$ cells. Times indicate duration of pre-culture of MEL cells with HMBA prior to inoculation (Marks et al 1978).

*Variant MEL cell lines*

Variant MEL cell lines have been developed from the parent cell line, DS19, which exhibit resistance or increased sensitivity to HMBA. Another variant MEL cell (V3.17) is resistant to the cytotoxic effects of 2–5 ng/ml vincristine, a relatively low level of drug resistance compared to several multidrug resistant cell lines which are 20-fold to several hundredfold more resistant (Melloni et al

1988). MEL cell line V3.17 compared to DS19 is: (1) more sensitive to HMBA, for example, one-tenth the concentration of HMBA optimal for inducing vincristine-sensitive DS19 cells will induce over 50% of V3.17 cells; (2) induced to commit with little or no latent period (Fig. 2C); (3) more rapidly induced to accumulate benzidine-reactive cells and to commit to terminal cell division; and (4) resistant to the inhibitory effects of dexamethasone on HMBA-induced differentiation. Several independently derived MEL cell lines resistant to vincristine have similar characteristics. These findings suggest that the mechanism for vincristine resistance in these cells develops in association with the mechanism which increases the sensitivity to HMBA and eliminates the latent period for induced differentiation. The resistance of V3.17 cells to the inhibitory effects of dexamethasone is consistent with an action of this inhibitor at a late step during the latent period. The lack of the latent period during HMBA-induced differentiation of V3.17 cells appears to reflect constitutive expression of a factor which circumvents HMBA-induced early events required for expression of differentiation. This factor (or factors) could be the active product of critical latent period events.

A cell line, R1, was developed from MEL cell line DS19 which is totally resistant to HMBA-induced differentiation and suppression of tumorigenicity (Marks et al 1983). When we developed a variant of R1 cells which is vincristine resistant, designated R1vcr, (J Michaeli, RA Rifkind and PA Marks, unpublished observations), the HMBA-resistant phenotype of R1 cells was overcome. Indeed, R1vcr cells have a phenotype similar to that of V3.17 cells, namely, a marked increase in sensitivity to HMBA and little or no latent period for induced differentiation and loss of proliferative capacity.

What about the mechanism of this altered phenotype associated with vincristine resistance? The finding that a number of independently derived vincristine-resistant MEL cell lines all show this same increased HMBA-responsive phenotype indicates that the mechanism of vincristine resistance is related to the mechanism of polar compound induction of differentiation and suppression of tumorigenicity. The mechanism of vincristine resistance in V3.17 cells appears to be related to that of the multidrug-resistant phenotype and may involve an energy-dependent export pump which reduces intracellular levels of certain agents (Riordan & Ling 1985). Evidence for this conclusion includes the following characteristics of multidrug-resistant cells: (1) accumulation of $^{3}$H-vincristine in V3.17 cells is substantially lower than that in DS19 cells; (2) V3.17 cells are cross resistant to colchicine; and (3) verapamil and nifedipine, agents which reverse vincristine resistance, enhance vincristine's cytotoxic effects on V3.17 cells. Although *mdr* mRNA (Riordan & Ling 1985) levels are increased in certain but not all multidrug-resistant cell lines, we have not yet detected increased amounts of *mdr* mRNA or the P-glycoprotein product of this gene, compared to DS19 cells (unpublished observations).

The evidence that vincristine resistance in these MEL cell lines has characteristics of the so-called multidrug-resistant phenotype makes it likely that a site of action of HMBA and related polar compounds is at the level of the membrane, possibly involving an altered membrane transport channel. HMBA action at the postulated membrane site could lead to a series of steps that cause expression of genes for products characteristic of the differentiated cell, including cessation of proliferation and suppression of tumorigenicity.

*Harris:* Have clinical trials been initiated?

*Marks:* HMBA is in Phase I clinical trials. In a series of 33 patients, there have been six objective responses involving partial remission of the cancers (Young et al, in press).

*Harris:* So this compound is not toxic at the appropriate dose?

*Marks:* Some patients tolerate it better than others. Patients who tolerate repeated prolonged exposure to HMBA are among those in whom an objective response has been observed. We need a better agent than HMBA, one that is biologically active in patients at lower concentrations than HMBA and shows little or no toxicity.

## References


Chen Z, Banks J, Rifkind RA, Marks PA 1982 Inducer-mediated commitment of murine erythroleukemia cells to differentiation: a multistep process. Proc Natl Acad Sci USA 79:471-475

Clarke MF, Kukowska-Latallo TF, Wesin E, Smith M, Prochownik EV 1988 Constitutive expression of a c-*myb* cDNA blocks Friend murine erythroleukemia cell differentiation. Mol Cell Biol 8:884-892

Fibach E, Reuben RC, Rifkind RA, Marks PA 1977 Effect of hexamethylene bisacetamide on the commitment to differentiation of murine erythroleukemia cells. Cancer Res 37:440-444

Friend C, Scher W, Holland J, Sato T 1971 Hemoglobin synthesis in murine erythroleukemia cells *in vitro*: stimulation of erythroid differentiation by dimethylsulfoxide. Proc Natl Acad Sci USA 68:378-382

Gusella JF, Geller R, Clarke B, Weeks V, Housman D 1976 Commitment to erythroid differentiation by Friend erythroleukemia cells: a stochastic analysis. Cell 9:221-229

Marks PA, Rifkind RA 1984 Differentiation modifiers. Cancer 54:2766-2769

Marks PA, Reuben R, Epner ET et al 1978 Induction of murine erythroleukemia cells to differentiate: a model for the detection of new anti-tumor drugs. Antibiot Chemother 23:33-41

Marks PA, Chen ZS, Banks J, Rifkind RA 1983 Erythroleukemia cells: variants inducible for hemoglobin synthesis without commitment to terminal cell division. Proc Natl Acad Sci USA 80:2281-2284

Marks PA, Sheffery M, Rifkind RA 1987 Induction of transformed cells to terminal differentiation and the modulation of gene expression. Cancer Res 47:659-666

Melloni E, Pontremoli S, Damiani G et al 1988 Vincristine-resistant erythroleukemia cells have marked increased sensitivity to hexamethylene bisacetamide induced differentiation. Proc Natl Acad Sci USA 85:3835-3839

Ramsay RG, Ikeda K, Rifkind RA, Marks PA 1986 Changes in gene expression associated with induced differentiation of erythroleukemia: proto-oncogenes, globin genes and cell division. Proc Natl Acad Sci USA 83:6849-6853

Reuben RC, Wife RL, Breslow R, Rifkind RA, Marks PA 1976 A new group of potent inducers of differentiation in murine erythroleukemia cells. Proc Natl Acad Sci USA 73:862-866

Reuben R, Khanna PL, Gazitt Y, Breslow R, Rifkind RA, Marks PA 1978 Inducers of erythroleukemic differentiation: relationship of structure to activity among planar-polar compounds. J Biol Chem 253:4214-4218

Riordan JR, Ling V 1985 Genetic and biochemical characterization of multidrug resistance. Pharmacol & Ther 28:51-57

Tanaka M, Levy J, Terada M, Breslow R, Rifkind RA, Marks PA 1975 Induction of erythroid differentiation in murine virus infected erythroleukemia cells by highly polar compounds. Proc Natl Acad Sci USA 72:1003-1006

Tsiftsoglou AS, Robinson SH 1985 Differentiation of leukemia cell lines: a review focusing on murine erythroleukemia and human HL-60 cells. Int J Cell Cloning 3:349-366

Young CW, Fanucchi MP, Walsh TD et al 1988 Phase I trial and clinical pharmacological evaluation of hexamethylene bisacetamide administration by 10-day continuous intravenous infusion at 28 day intervals. Cancer Res 48:7304–7309

# Loss of genetic information in cancer

Webster K. Cavenee, Marc F. Hansen, Heidi J. Scrable and C. David James

*Ludwig Institute for Cancer Research, Royal Victoria Hospital, 687 Pine Avenue West, Montreal, Quebec H3A 1A1, Canada*

*Abstract.* The determination and comparison of genotypic combinations at genomic loci in normal and tumour tissues from patients with various types of cancer have defined the chromosomal locations of loci at which recessive mutations play a role in disease. The predisposing nature of some of these mutant alleles is exemplified in studies of retinoblastoma and osteogenic sarcoma. These two clinically associated diseases share a pathogenetically causal predisposition that maps to chromosome position 13q14. A similar mechanism at 11p15.5 is involved in the development of the embryonal variant of rhabdomyosarcoma, Wilms' tumour and hepatoblastoma. Finally, genomic alteration of chromosome 10 is apparent in glioblastomas and mixed tumours of glioblastoma/astrocytoma grade III but not in homogenous astrocytoma grades II or III, suggesting the definition of a locus involved in tumour progression and, perhaps, an approach to molecular genetic staging of tumours.

*1989 Genetic analysis of tumour suppression. Wiley, Chichester (Ciba Foundation Symposium 142) p 79–92*

The suggestion that cancer might have a genetic aetiology was first made based on abnormalities in the mitotic cycle of sea urchin eggs which were frequent precursors of aberrations in growth and chromosomal content. These observations have led to analytical approaches to cancer genetics and the development of experimentally malleable systems to uncover the underlying basis for such behaviour.

There is now a great deal of information which supports this approach: a) the phenomenon of clinical aggregation of tumours of specific histological types in families has been extensively described; b) specific alterations of the germline chromosomes of some individuals seem to confer the propensity to develop particular tumours; c) the availability of increasingly precise methods for detection of genomic aberrations have allowed the identification of somatically acquired rearrangements in tumours; d) individuals with defects in repair of genotoxic damage exhibit greatly increased risk for tumour development; and finally, e) the ability of chemical and physical agents to mutate DNA appears to be mirrored by their capacity to elicit tumours in animals. These observations are reviewed more fully in Knudson (1986) and Hansen & Cavenee (1987). In this chapter, we review the work of our laboratory that bears on several of these questions and which provides evidence that the loss of genetic information is a frequent aetiological mechanism in predisposition to, and progression of, human cancers.

## Predisposition

Among the inherited conditions that appear to predispose to cancer are recessively transmitted perturbations of the immune system, defective DNA repair and abnormal sexual development. The direct relationship between these impaired functions and increased risk of cancer in such individuals is reasonably certain. Another group of conditions predisposing to neoplasia displays familial patterns consistent with the inheritance of an autosomal dominant trait(s). About 50 different types of human neoplasia have been described in terms of familial aggregation (Mulvihill 1977) and each of these genetic cancers has corresponding sporadic forms in the general population. Additionally, breast cancer, colon cancer and ovarian tumours are frequently individually clustered in families. However, in some cases, the occurrence of such tumours may be a partial manifestation of the 'Cancer family syndrome' in which the clustering of more than one specific type of cancer is apparent. Each of these tumours has also been reported to cluster in families, inherited as an autosomal dominant trait. Family studies suggest that several inherited genetic defects may predispose to related but clinically distinct tumours. Thus, at least five different inherited forms of breast cancer have been distinguished and 5% of all breast cancer patients are members of such families. Similarly, inherited colon polyposis and colon cancer fall into more than 10 different groups.

The interpretation of the genetic transmission of cancer in these families as single dominant mutations is, however, at odds with three lines of evidence. First, if a single mutation were sufficient to elicit a tumour, then families segregating for autosomal dominant forms of cancer would be expected to have no normal tissue in the diseased organ. This expectation is in direct contrast to the clinical observation of discrete tumour foci amidst normal, functional tissue in such individuals. Secondly, elegant epidemiological analyses (Knudson 1986) of sporadic and familial forms of several cancer types have indicated that the conversion of a normal cell to a tumour cell requires multiple events. Finally, there is a substantial body of evidence derived from somatic cell hybrids which indicates dominance of the wild-type phenotype in the presence of tumorigenic mutations (Stanbridge 1987, Klein 1987).

Our work for the past few years has been directed at understanding this paradox. The first studies were of retinoblastoma, a neoplasm of embryonic retina which occurs in heritable and sporadic forms. Statistical analysis of various characteristics of these disease forms (Knudson 1971) has led to the model in Fig. 1. This model requires as few as two mutations to effect malignant transformation. The hereditary cases would have inherited a ger-

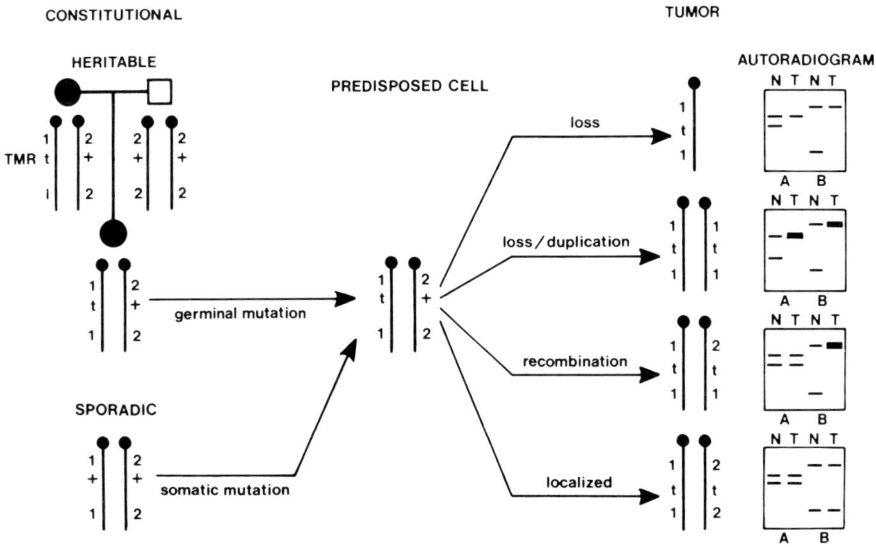


FIG. 1. A model for chromosomal mechanisms that could result in nullizygosity for a tumour locus. In an hereditary case (upper left side) the child inherits a chromosome from her affected mother which carries a recessive defect at the *TMR* locus (designated 't') and so is genetically t/+ in all her cells. A tumour would occur after the elimination of the dominant wild-type allele at the *TMR* locus by one of the mechanisms shown. In sporadic cases (lower left side) a recessive mutation could lead to a tumour when it is unmasked by one of the mechanisms. 1 and 2 designate allelic variants at arbitrary loci on the chromosome carrying the *TMR* locus.

minal mutation, which does not itself cause the tumour but rather predisposes each retinal cell to a further transforming event. In this model, the non-hereditary cases would also result from two mutations except that these events would have to occur in the same somatic cell. Thus, the two forms of the disease could be viewed as resulting from the same two-step process at the level of the aberrant retinal cell, the difference being the inheritance or somatic occurrence of the first mutation.

We proposed (Cavenee et al 1983) that the second step in tumorigenesis in both heritable and sporadic retinoblastoma involves somatic alteration of the normal allele at the 'tumour locus' in a way that unmasks the mutant allele. Thus, the first mutation in this process, although it may be inherited as an autosomal dominant trait at the level of the organism would be expected to have the properties of a recessive defect in the individual retinal cell. In this model, the heritable form of the disease arises as a germinal mutation of the *TMR* locus and is inherited by an individual who, therefore, is an obligate heterozygote (t/+) at the *TMR* locus in each of their somatic and germ cells. A subsequent event in any retinal cell which results in homozygosity for the mutant allele (that is, mutant at the *TMR* locus on both chromosome homologues) will result in a tumour clone. The chromosomal mechanisms that could accomplish this loss of heterozygosity include: mitotic non-disjunction with loss of the wild-type chromosome, which would give hemizygosity at all loci on the chromosome; mitotic non-disjunction with duplication of the mutant chromosome, which results in homozygosity at all loci on the chromosome; or mitotic recombination between the chromosomal homologues with a breakpoint between the *TMR* locus and the centromere, which would cause heterozygosity at loci in the proximal region and homozygosity throughout the rest of the chromosome, including the *TMR* locus. Regional events such as gene conversion, deletion or mutation must also be considered. Heritable and sporadic retinoblastoma could each arise through the appearance of homozygosity at the *TMR* locus, the difference being two somatic events in the sporadic case, as compared to one germinal and one somatic event in the heritable case.

Our approach is schematically illustrated on the right side of Fig. 1. We have compared alleles at loci on various chromosomes in tumour and normal tissues from individuals; such loci are defined by restriction fragment length polymorphisms. It is apparent from the diagram that each of the somatic chromosomal changes proposed has distinct experimental predictions. An example of the results obtained is shown for the case of Rb-409 in Fig. 2. The normal cells of this patient were heterozygous at three loci on chromosome 13 which lie towards the centromere (*HU10*), the telomere (*9A7*) and the middle of the chromosome (*HU26*). In the corresponding tumour tissue, there was only one apparent allele at each locus and, in each case, the dosage of the allele was increased with respect to its normal counterpart. These findings are consistent with the loss/duplication mechanism depicted in Fig. 1. We have examined a large series of cases of retinoblastoma in this way and this relatively gross mechanism appears to be involved in about 75%. Furthermore, the idea that these chromosomal gymnastics uncover predisposing cellular recessive mutations was strongly supported by the demonstration that in familial cases the chromosome remaining in the tumour was inherited from the affected parent (Cavenee et al 1985).

These studies suggested the possibility of identifying loci which predispose to other types of tumours by examining the incidence of loss of heterozygosity at loci on each arm of each chromosome. Some of our attempts to test this proposal are represented in Table 1. We first searched for somatic loss of heterozygosity in the embryonal nephroblastoma, Wilms' tumour (Koufos et al 1984). This tumour, the most common abdominal neoplasm in children, exhibits both heritable and sporadic forms. The autosomal dominant heritable form is often associated with other specific congenital malformations, such as aniridia, genito-urinary abnormalities and mental retardation. A chromosomal locus which plays a role in the development of this tumour has

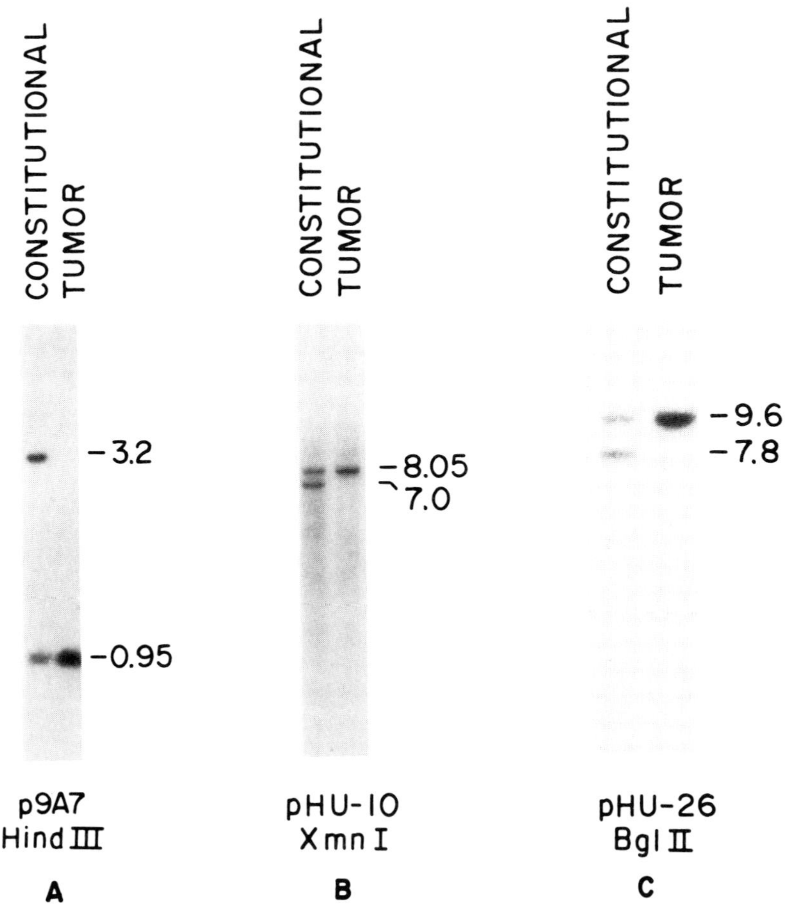


FIG. 2. Homozygosity effected by segregation of one chromosome 13 homologue with duplication of the remaining one. Southern blots of (A) *Hin*dIII-digested DNA hybridized to p9A7 which contains an insert homologous to a locus on chromosome 13 that maps to band q31; (B) *Xmn*I-digested DNA hybridized to the insert from plasmid pHU10 which is homologous to a locus on chromosome 13 that maps to q13; (C) *Bgl*III-digested DNA hybridized to the insert from plasmid pHU26 which is homologous to a locus on chromosome 13 that maps to band q21.

been mapped to chromosome 11 band p13 by the identification of interstitial deletions of 11p13 in germline karyotypes of patients, and by direct cytogenetic determination of 11p13 deletions in Wilms' tumour cells in patients with normal germline karyotypes.

We examined normal and tumour tissue from seven children with Wilms' tumour at loci on the short arm of chromosome 11. In five of these cases, heterozygosity was lost in the tumour, similar to the results shown in Table

**TABLE 1 Loss of heterozygosity for chromosome 11p loci in clinically associated tumours**

| | | *Alleles present* | | | | | |
|---|---|---|---|---|---|---|---|
| | | *pADJ-762* *(MspI)* | *(BclI)* | *pHins-310* *(PvuII)* | *JW151* *(HindIII)* | | *pTBB2* *(TaqI)* |
| *Patient* | *Tissue* | | | | $\gamma^G$ | $\gamma^A$ | |
| A. Wilms' tumour | | | | | | | |
| 1 | N | 2,2 | 1,2 | 1,2 | 1,2 | 1,2 | 2,2 |
| | T | 2,2 | 2,2 | 2,2 | 1,1 | 1,1 | 2,2 |
| 2 | N | 1,2 | 2,2 | 1,2 | 1,1 | 1,1 | 1,2 |
| | T | 2,2 | 2,2 | 2,2 | 1,1 | 1,1 | 2,2 |
| 3 | N | 1,2 | 1,1 | 1,3 | 1,2 | 1,1 | 2,2 |
| | T | 2,2 | 1,1 | 3,3 | 1,1 | 1,1 | 2,2 |
| B. Hepatoblastoma | | | | | | | |
| 1 | N | 1,2 | 1,1 | 2,2 | 1,1 | 1,1 | 1,2 |
| | T | 2,2 | 1,1 | 2,2 | 1,1 | 1,1 | 1,1 |
| 2 | N | 2,2 | 1,2 | 1,3 | 2,2 | 1,2 | 2,2 |
| | T | 2,2 | 2,2 | 1,1 | 2,2 | 1,1 | 2,2 |
| C. Rhabdomyosarcoma | | | | | | | |
| 1 | N | 1,2 | — | 1,2 | — | — | 1,1 |
| | T | 1,1 | — | 2,2 | — | — | 1,1 |
| 2 | N | 2,2 | — | 1,2 | — | — | 1,2 |
| | T | 2,2 | — | 1,1 | — | — | 1,1 |

—, not tested; N, normal; T, tumour.

1A. Furthermore, we also became aware of the clinical association of Wilms' tumour with the other embryonal tumours, rhabdomyosarcoma and hepatoblastoma. A common aetiology for these tumours is suggested by their association with the autosomal dominant, Beckwith-Wiedemann syndrome, that comprises multiple congenital malformations and concurrent high risk for the development of specific rare childhood tumours. More than 10% of all individuals with Beckwith-Wiedemann syndrome develop rare cancers, including Wilms' tumour, hepatoblastoma, rhabdomyosarcoma and adrenal carcinoma, in association with the disorders of excessive growth that characterize the syndrome. Also notable is the ability of these embryonic tumours to undergo divergent differentiation to form a tissue that does not occur in the normal end-stage organ. For example, fetal rhabdomyomatous nephroblastoma is a variant Wilms' tumour which is composed of striated muscle of the kidney. The clinical association of Wilms' tumour with other specific rare tumours in the presence of the growth excess syndromes, the development of more than one rare tumour in the same individual, and the presence of

**TABLE 2 Losses of heterozygosity in human tumours**

| *Chromosome region* | *Tumour type* |
|---|---|
| 1p | Neuroblastoma |
| 3p | Renal cell carcinoma, small cell lung carcinoma |
| 5q | Colorectal carcinoma |
| 10 | Glioblastoma |
| 11p | Wilms' tumour, hepatoblastoma, rhabdomyosarcoma, breast carcinoma, adrenal cortical carcinoma, transitional cell bladder carcinoma |
| 13q | Retinoblastoma, osteosarcoma, breast carcinoma |
| 17 | Astrocytoma, colorectal carcinoma |
| 18 | Colorectal carcinoma |
| 22 | Meningioma, acoustic neuroma |

heterotropic tissue in these tumours could be simply coincidental. Alternatively, these clinical associations could reflect a common aetiological event and each of the developmental anomalies could arise from mutation of the same locus. Such mutations could be revealed by mitotic segregation events, similar to those demonstrated for Wilms' tumour and retinoblastoma, which would specifically produce hepatoblastoma and rhabdomyosarcoma tumours that have lost heterozygosity. The experimental test of this hypothesis showed (Koufos et al 1985) that both rhabdomyosarcomas and hepatoblastomas specifically lost heterozygosity at loci on chromosome 11p, as shown in Table 1 B and C. This is consistent with the idea that the clinical association between these tumours and Wilms' tumour arises through a shared pathogenetic mechanism.

These results suggest a general approach to identifying the chromosomal positions of loci whose recessive alleles predispose to human cancers. Many laboratories have taken such an approach and the resulting information is summarized in Table 2. Several points provide some assurance about the applicability of the genetic model. First, there are enough of these loci to be reasonably certain that they are not general 'cancer genes'. Secondly, the chromosomal mechanisms originally identified in embryonal tumours also occur in adult neoplasia. Finally, the tumours listed include representatives of each histogenic derivation.

## Progression

The final aspect we would like to address is the utility of genotypic determinations in the delineation of stages of malignant evolution in neoplastic disease. Our initial studies (James et al 1988) in this regard are of tumours of

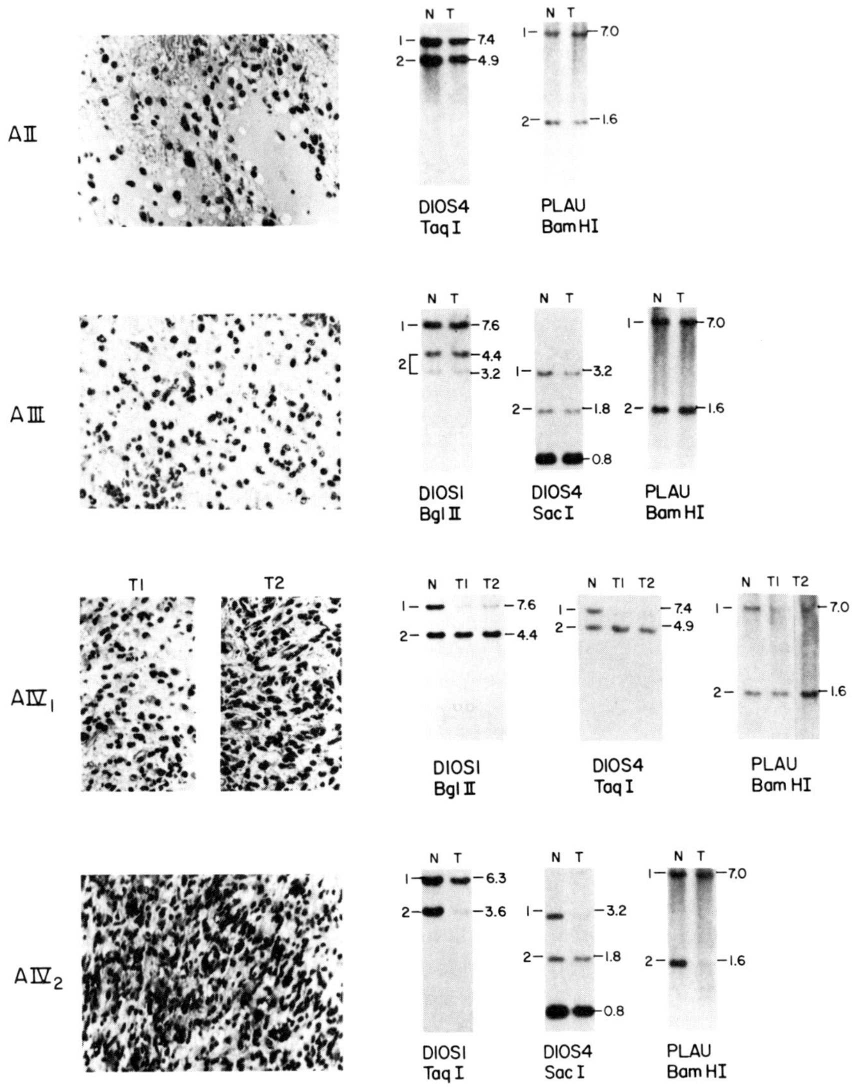


FIG. 3. Stage-specific loss of heterozygosity for loci on chromosome 10 in tumour tissue from astrocytomas. Histological grade and corresponding section(s) are shown on the left. Southern blots have the observed alleles labelled on the left with corresponding lengths in kb to the right. Loci examined and restriction enzymes used are indicated beneath each normal (N) – tumour (T) set.

astrocytic origin, the most frequently occurring neoplasms of the central nervous system. Identification of the astrocytic nature of a tumour depends upon conventional histopathology in combination with histochemistry or immunohistochemistry. The histopathological characteristics indicating biological aggressiveness or malignancy in these tumours have been empirically determined and the tumours are now grouped in four malignancy grades. Prognoses vary accordingly, with five-year post-operative survival rates reaching a nadir of less than 5% for patients whose tumours are diagnosed as grade IV astrocytomas (also referred to as glioblastomas), even with treatment programmes following surgery which involve radiation and/or chemotherapy. Clinical recurrence of astrocytoma is frequent and histological examination commonly reveals that the recurrent tumour is less well differentiated than its predecessor — observations which suggest that astrocytoma is a progressive disease.

We compared normal and tumour genotypes at loci on chromosome 10 for DNA samples from 26 adult cases of astrocytoma, histologically representing a continuum of malignancy grades. Allelic combinations were determined at loci homologous to three different chromosome 10 probes: representative cases are shown in Fig. 3. Each of 15 high grade (AIV) tumours showed loss of heterozygosity at one or more loci. In sharp distinction, none of 11 lower grade (AII or AIII) tumours showed a loss of alleles at any of the loci. Several of the grade IV astrocytomas analysed displayed areas of varying cellularity and diverse morphology with or without necrosis. In three cases, such areas were dissected separately and DNA isolated for restriction fragment length polymorphism analysis. In each instance, DNA from these different areas showed loss of heterozygosity for chromosome 10 loci consistent with the high malignancy grade classification of the tumour (Fig. 3: $AIV_1$).

These results are important in several ways. First, they suggest that while morphological and karyotypic variation may exist within the cells constituting a glioblastoma, all of the different cell subpopulations would have evolved from a common, karyotypically abnormal precursor. Secondly, the clinical/histopathological evidence that glial tumours usually recur and that recurrent tumours frequently display a malignant progression suggests the existence of a molecular pathway or pathways along which gliomas evolve with increasingly malignant potentials. Thirdly, they are the beginning of a genotypic, rather than phenotypic, approach to defining the molecular basis of tumour progression. Finally, they suggest that some of the genetic events summarized in Table 2 may not be involved with predisposition but rather with malignant progression and so emphasize the complementarity of formal genetics and tumour biology.

## References

Cavenee WK, Dryja TP, Phillips RA et al 1983 Expression of recessive alleles by chromosomal mechanisms in retinoblastoma. Nature (Lond) 305:779–784

Cavenee WK, Hansen MF, Kock E et al 1985 Genetic origins of mutations predisposing to retinoblastoma. Science (Wash DC) 228:501–503

Hansen MF, Cavenee WK 1987 Genetics of cancer predisposition. Cancer Res 47:5518–5527
James CD, Carlbom E, Dumanski J et al 1988 Genomic alterations in glioma malignancy stages. Cancer Res 48:5546–5551
Klein G 1987 The approaching era of the tumor suppressor genes. Science (Wash DC) 238:1539–1545
Knudson AG 1971 Mutation and cancer: statistical study of retinoblastoma. Proc Natl Acad Sci USA 68:820–823
Knudson AG 1986 Genetics of human cancer. Annu Rev Genet 20:231–251
Koufos A, Hansen MF, Lampkin BC et al 1984 Loss of alleles at loci on human chromosome 11 during genesis of Wilms' tumour. Nature (Lond) 309:170–172
Koufos A, Hansen MF, Copeland NG, Jenkins NA, Lampkin BC, Cavenee WK 1985 Loss of heterozygosity in three embryonal tumours suggests a common pathogenetic mechanism. Nature (Lond) 316:330–334
Mulvihill JJ 1977 Genetic repretory of human neoplasia. In: Mulvihill JJ et al (eds) Genetics of human cancer. Raven Press, New York p 137–143
Stanbridge EJ 1987 Genetic regulation of tumorigenic expression in somatic cell hybrids. Adv Viral Oncol 6:83–101

## DISCUSSION

*Benedict:* I think we should mention the differences in chromosome 11 between the Wilms' tumour locus and the rhabdomyosarcoma locus. The embryonic rhabdomyosarcoma locus appears to be in band p15 and, although the Beckwith-Wiedemann syndrome includes the Wilms' tumour, the Wilms' tumour locus is in p13.

*Cavenee:* That's not true. Wilms' tumour in the genetic form is not, as far as we can tell in families, in chromosome 11p13. Linkage analysis in two families that we have and a few families that Louise Strong has shows no evidence for linkage anywhere on the short arm of chromosome 11. That suggests that even though we are seeing loss of heterozygosity for chromosome 11 in sporadic Wilms' tumours, that ascertainment of predisposition isn't necessarily holding in familial cases. I don't know why that should be so, it could be selection for non-lethal mutations in the genesis of Wilms' tumour families.

Wilms' tumours are not homogeneous: there are biological and molecular distinctions. One type of Wilms' tumour does, by this analysis, seem to reside in 11p13; another type resides in 11p15—that's the type which is associated with hemihypertrophy and the other tumours as well.

*Stanbridge:* Do you wish to make any comments regarding the *Rb* locus in this type of analysis?

*Cavenee:* We have two pieces of conflicting evidence. One is that we rarely see either genomic or transcriptional alterations in primary retinoblastoma or osteogenic sarcomas. The second is, we have looked at children with germline translocations who have developed retinoblastoma for the positioning of the translocation breakpoints. Those breakpoints cluster in one of the introns of the *Rb* gene which has been isolated by Bob Weinberg. That, I think, supports the identity of that gene as the basis of retinoblastoma.

*Stanbridge:* Always within the same intron?

*Cavenee:* Yes, it's a big intron, about 180 kb. We don't have any markers in that intron but this is in unrelated people.

*Harris:* So you envisage the possibility that progression in a tumour may be due to a loss of function succeeded by further losses of function?

*Cavenee:* I hope so.

*Harris:* I think it's highly probable, because mutations conferring a gain of function must be extraordinarily rare.

*Cavenee:* The caveat on that progressive change is that what we are looking at is necessarily half dose or less. In this case we could be looking at gene dosage; mutation may not even be necessary.

*Weinberg:* Have you looked at the *Rb* locus in any of these rhabdomyosarcomas? Was it ever involved?

*Cavenee:* It was never involved as examined at the level of genomic rearrangement. Transcripts were also present at normal levels, in those cases where we could look at them.

*Klinger:* Has the site of the rearrangement on chromosome 11 in transitional cell carcinoma of the bladder been localized?

*Cavenee:* Not that I know of.

*Klinger:* Do you have an estimate for how many distinct putative suppressor loci have been identified on chromosome 11?

*Cavenee:* We have two at this level; we could have a whole linkage set, which I would guess is probably the case.

*Noble:* What about interactions with the positively acting oncogenes? Do any of the oncogenes segregate in an interesting way with any of the losses of expression?

*Cavenee:* The only one which is in a position to be interesting in that context is Harvey-*ras*, which is on the short arm of chromosome 11. I don't *think* that has much to do with anything. There don't seem to be any alterations of Harvey-*ras* transcripts or proteins in those tumours.

*Noble:* But one sees *erbB* amplification in many of the retinoblastomas.

*Cavenee:* That also occurs in cell lines and in 30 or 40% of glioblastomas, mainly in the advanced stages. *erbB* amplification changes between primary tissue and and tissue culture, it disappears in some of the latter, as far as we can tell.

*Balmain:* We have been looking in animal model systems at deletions and loss of heterozygosity of mouse chromosomes during tumour progression. In a very high proportion of skin carcinomas which have an activated Harvey-*ras* gene, we find that the normal H-*ras* gene is lost or under-represented, either by loss of the whole mouse chromosome 7 or by loss of parts of that chromosome by mitotic recombination. We have looked at about 10 histologically similar tumours which do not have activated *ras* genes, and we haven't found loss of chromosome 7 in any of those. In this model there is a very close association between activation of the Harvey-*ras* gene and loss of normal *ras* alleles on the other chromosome.

*Harris:* In these cases where you eliminate a whole chromosome, is the cell otherwise diploid or is it aneuploid?

*Cavenee:* In glioblastomas the cells are certainly aneuploid.

*Harris:* The reason I ask is that 20 years ago, we often found in mouse tumours or tumours produced by hybrid cells a situation like the one you describe. One of the chromosomes of the tumour cell is duplicated and one of the normal homologues is lost. So you think you have the normal homologue but in fact you haven't. I thought then that the doubling up of one of the chromosomes of the tumour cell had to precede the elimination of the normal homologue. I don't know whether anybody has cells that are monosomic for an autosome, but otherwise euploid. They may exist, but they must be very rare.

*Bodmer:* We have looked at cell lines in colorectal carcinomas. Unfortunately, in those cases we don't have the normal tissue, but because of the very informative nature of Alec Jeffrey's DNA probes we can be reasonably sure that in the majority of cases where we see only one allele there is allele loss. Some colorectal carcinomas are remarkably near diploid, perhaps about half of them, and we know in those cases that when you have allele loss you do see two chromosomes 5. I don't think that agrees with the allele loss occurring before the duplication. I think in most cases you get monosomy: monosomy is disadvantageous at the cellular level, so there is rapid reduplication to restore disomy because of the advantage of gene balance for genes on the chromosome other than the one whose mutation you are selecting.

*Harris:* That's possible. I haven't examined the question systematically.

*Bodmer:* We know that the mutations come first, at least in these colorectal carcinomas.

*Benedict:* We found in primary retinoblastoma that about 25% of those cases are monosomic for chromosome 13. However, there was one case with monosomy for chromosome 13 that we passaged in the nude mouse just once. It very rapidly became diploid for chromosome 13, both in culture and *in vivo*. So the model system showed rapid strong selection against monosomy. And it does say that the monosomy came first.

*Stanbridge:* Lloyd Old's group studying human melanomas showed loss of isoenzyme and RFLP polymorphism in melanoma cell lines at many different loci (Dracopoli et al 1985). One has to be careful when interpreting data derived from cultured cells versus the primary tumours or the metastatic tumours.

*Klein:* I would like to ask about somatic crossing over. According to classical dogma, it requires somatic pairing, like in the *Drosophila* tunia spot system.

This was thought not to occur in vertebrates. You showed previously that it did occur in retinoblastoma but it was only a fraction of all these second gene losses. Now you are saying that it is quite frequent. Does that mean there is a different frequency of somatic pairing in different tissues?

*Cavenee:* No, I don't think so. In retinoblastomas the markers that we had were a quarter of the way down the chromosome, so a mitotic recombination above that would not have been detected and would have been called duplication.

*Klein:* Is your idea then that mitotic crossing-over is a frequent phenomenon in all tissues in vertebrates?

*Cavenee:* Yes, I think so. It has been detected experimentally in Chinese hamster ovary cells where there are more chromosome markers, including selectable markers.

*Sporn:* You showed an interesting tumour that had a mixed low-grade astrocytic morphology as well as a high-grade glioblastoma morphology. Do you have any idea what is the difference between these two pieces of tissue? They both have lost chromosome 10.

*Cavenee:* The only thing is that in two of the tumours that we have looked at, which may have been the only two that are pairs, there was a mixture of glioblastoma with either second or third grade astrocytoma. That's all based on the number of nuclei. We have seen differences, not in chromosome 10, but in amplification of epidermal growth factor receptor.

*Sporn:* Is it more heavily amplified in the glioblastoma?

*Cavenee:* Yes. Glioblastomas seem to be the end stage of brain tumours in general.

*Noble:* That's right, and it often seems that everything that gets very malignant is called a glioblastoma. Although the morphological identification of all tumours presents problems, in brain tumours the standards are even cruder. This is in part because of our lack of knowledge about the normal lineages of the central nervous system (CNS). We believe that some of the glioma designations will turn out to be incorrect, and will eventually be replaced by more stringent cellular biological designations.

*Tarin:* I disagree, for example, an oligodendroglioma is very easy to recognize. It is an unusual and rare tumour with several distinguishing features, such as the boxing of the nuclei and the monotonous appearance of the small cells constituting the tumour. The astrocytoma too is, in most cases, easy to diagnose. Nor would I accept that the glioblastoma is necessarily the end pathway of all neural tumours. Thus, although histological classification has its limitations, sharp discrimination between different tumour types with different behaviour and prognosis is usually possible.

*Noble:* Despite this supposed ease of diagnosis, when we have looked at antigen expression, response to mitogens and ability to stimulate endothelial cell proliferation (three pretty fundamental aspects of tumour cell biology) within, for example, oligodendrogliomas, we can recognize two distinct tumour groups. In fact, we can see these two classes in every type of glioma. There are glioma-derived populations that express central nervous system antigens, and populations derived from other morphologically identical gliomas which do not. These two antigenically defined categories differ from each other in their response to epidermal growth factor and fibroblast growth factor, in their expression of N-CAM (the neural cell adhesion molecule), and in the promotion of endothelial proliferation. Moreover, all gliomas within an antigenically defined category express these properties similarly, regardless of the classification assigned to them by neuropathologists.

*Tarin:* When you are looking at the level of individual molecules it is quite possible that different cell populations in the brain may share certain markers, for example, tissue-specific proteins, and might respond to the same kind of mitogens. However, there is no doubt that when you take large numbers of CNS tumours, or any other tumours for that matter, and identify them by histological criteria and then follow their prognosis, they form fairly well recognizable groups. Clinical behaviour correlates quite well with the diagnosis under a microscope and one cannot, therefore, abandon the value of separating tumours into different categories simply because they sometimes share common antigenic markers.

## Reference

Dracopoli NC, Houghton AN 1985 Loss of polymorphic restriction fragments in malignant melanoma, implications for tumor heterogeneity. Proc Natl Acad Sci USA 82:1470–1474

# General discussion II

## Genetics of polyposis and colorectal cancer

*Bodmer:* Familial adenomatous polyposis (FAP) is a disease in which very many polyps occur throughout the colon and also, to a lesser extent, in the upper gastrointestinal tract. Patients with FAP quite often have other tumours, such as desmoids. This has led to the suggestion that the disease may be heterogeneous.

Herrera et al (1986) described an individual with multiple abnormalities and polyposis who had a band missing from the middle of 5q. We presumed, because that individual had multiple developmental abnormalities together with polyposis, there was a reasonable chance that the FAP gene was in that missing band and that it fitted the classic retinoblastoma model. We were lucky enough to be given a probe called C11p11 by Bob Williamson, that showed very close linkage (Bodmer et al 1987). There is now another probe on chromosome 5 called π227, which is more informative. (The level of heterozygosity for C11p11 is only about 15% in the populations we have studied. We now believe that it is more frequent in populations of Celtic origin than others.) The mapping data are, in our view, consistent with FAP being in between π227 and C11p11 with about 4% recombination on either side. The key to further studies is to identify more individuals who have the 5q deletion.

The first individual that we found with that deletion is a 12-year-old child who has not yet developed polyposis (E. Solomon, personal communication). *In situ* annealing by Denise Sheer placed the C11p11 marker around the 5q21 band, showing that neither C11p11 nor π227 appear to be lost in that deletion.

We are collaborating with an Australian group that has a family in which that deletion is segregating. The mother had the deletion and polyposis; two of her children also have the deletion and polyposis (KA Hockey, MT Mulcahy, P Montgomery, S Levitt, personal communication). This demonstrates the transmission of polyposis with the deletion through the family, and makes the loss theory entirely consistent with the other results on allele loss. We are now growing cell lines from these two children for use in molecular studies.

We see no evidence of heterogeneity of linkage in the clear examples of polyposis. Thus, we see no evidence that families with and without the associated Gardner's type phenomena show different linkage with C11p11 or π227. There are, however, other families with dominantly inherited colorectal carcinoma that don't have many polyps, and there are some examples with very late development of polyps. These could well be due to different genes, but the majority class, including what has been called Gardner's, are, we think, all mutations of the same gene.

Ellen Solomon in collaboration with us and others has been looking at allele loss in sporadic colon carcinomas (Solomon et al 1987). We showed allele loss initially through a marker λm58 obtained from Alec Jeffreys that is on the tip of chromosome band 5q. About 20 to 30% of sporadic carcinomas showed allele loss with this marker. Ellen Solomon has now accumulated more data using other markers, including one that is on the short arm. By using these markers along the chromosome you can sometimes exclude mitotic recombination and be sure that there is monosomy at some stage, implying loss of a whole chromosome.

Initially, we thought that whole chromosome loss was the most common event. It is relatively common, but it has become clear that either mitotic events or interstitial events are more common than we at first thought. To my knowledge, there has been no sporadic carcinoma identified in which a 5q21 deletion has been proven. It is clear that amongst individuals with FAP, 5q21 deletion is a very rare event, more so than the equivalent deletion in retinoblastoma.

When you compare these different situations, including the renal cell carcinomas, there are clearly different frequencies of the different types of genetic events that are seen in the tumours. My belief is that the foremost explanation for that is selection, connected with the effect of losing a whole chromosome. If, for example, there is a cellular lethal gene on a chromosome, then that affects the viability of monosomics, and what parts of a chromosome can be lost. There may, of course, also be preferred sites at which different events may happen. The more complex the situation is with respect to whole chromosome loss versus other events, the higher the effective frequency of a chromosome 5 gene event is likely to be in the sporadic cases. Allele loss is at least 40% detectable with the markers that we have, so it is possible that most colorectal carcinomas have a change in that chromosome 5 FAP gene. Now that we can go back to paraffin blocks and get material suitable for Southern blots out of them we can do retrospective studies in situations where there is a lot of documented pathology. We have the impression that the frequency of allele loss in the tumours from colorectal carcinomas of patients with polyposis is lower than we would expect. If, for instance, the same events were involved and the frequency were the same in the tumours from the patients with polyposis as in the sporadic cases, that would indicate that 100% of the sporadic cases involved chromosome 5, because you know that 100% of the polyposis ones do. I think that's added evidence that we are getting to a situation where there is a very high frequency of the FAP-like genetic event in the sporadic colorectal tumours.

We have shown in one cell line from a FAP polyp and by looking at polyps directly that there is no evidence of allele loss in the polyps. Therefore, whatever else is happening genetically, there is no second genetic event in at least the majority of the polyps. Vogelstein's arguments for the clonality of polyps (Fearon et al 1987) are probably wrong, because he is looking at clonality of X-linked markers, where the 'patch' size from Bruce Ponder's data for the colon is relatively high. This means that X marker clonality is essentially irrelevant to what happens at the stage at which the carcinomas might develop.

It is an interesting problem, that polyps are a focal event and yet may not be clonal. There is a challenge to understanding how you can get focal events that are not clonal. If there is variation in the frequency with which a given cell produces a given level of product, then in the healthy individual, the distribution reflects the normal distribution of the FAP gene product. In an individual with FAP, the distribution curve is such that, on average, half as much product is made. If the level of this gene product in a normal cell is close to what is needed, when it is reduced by half, it may relatively often, by statistical fluctuation, fall below the threshold amount that is required to contain growth. This model has to be a bit more complicated to explain locally spread growth as, for example, in a polyp. Stochastically, however, you can get focal areas where, by chance, one or more cells produce too little of the product and that allows cellular outgrowth to produce a polyp in a focal way without it being clonal. Clonal development may then happen within the polyp.

With respect to other changes, Fearon et al (1987) have shown that chromosome 17p losses occur relatively frequently. They have been observed cytogenetically, and at least two cell lines have 17p monosomy as a result of translocations. Ellen Solomon has looked at 17p loss with similar results, namely about 30% loss of 17p markers. The initial results suggested there might be some association with loss of chromosome 5, but it now looks as though there is random distribution. Bos et al (1987) looked at Ki-*ras* in colorectal cancers. The Ki-*ras* change is a single dominant change, whereas the chromosome 5 and 17 changes are recessive. Again, there is so far no obvious evidence of an association between these various genetic events. (See also Vogelstein et al 1988.)

There are, therefore, a number of genetic changes now being found in colon carcinomas. In addition to those already mentioned, there is evidence for recessive changes on chromosome 18 and from Japan of changes on chromosome 22, loss of HLA class I antigens and ectopic expression of placental alkaline phosphatase (Monpezat et al 1988, Okamoto et al 1988). The functional relevance of these changes to tumorigenesis remains to be established.

All the studies so far on putative recessive effects involve looking at loss of *both* alleles and not yet (apart from retinoblastoma) looking at loss of the first allele. Although the polyps involve no allele loss in the cases of polyposis, I predict that in the sporadic tumours, where the first effect must be genetic, precancerous adenomas are clonal. We won't know that, however, until we have cloned the actual gene involved. All the measures of association between recessive genetic events are so far, therefore, very weak measures, because they look only at double events. If the gene dosage notion is accepted, the first loss could be important for the phenotype.

*Weinberg:* You are talking about an event you call focal but not clonal. It is my impression that it is both focal and clonal, and is also non-genetic. Are you confusing epigeneticity with clonality?

*Bodmer:* That's an interesting further distinction. I use focal advisedly. You can have non-clonal focal events that are the primary cause of a small local growth. There can be focal events that are clonal but the clonality is epigenetic. And there can be focal events that are clonal and the clonality is genetic. I believe that the focal event in polyposis is probably non-clonal. It may originally be clonal in the sense that this underproduction, and therefore tendency to grow, starts in one cell, but I believe that the release of growth factors leads to the growth of a number of surrounding cells—recruitment and a sort of autocatalytic effect.

*Benedict:* Are you saying that in some cases you will find two different karyotypic patterns under that scenario?

*Bodmer:* In the individual who has polyposis without a deletion, there is only the heterozygous defect in the polyposis gene. There is no obvious argument for any other genetic change.

*Benedict:* You just said that you didn't find any allele loss in polyps.

*Bodmer:* No allele loss means the second event hasn't occurred. The first event has—that's what is inherited in the family.

*Benedict:* Is that inherited first defect not detectable cytogenetically? Any solid tumour that I have ever examined is aneuploid and you should be able to detect those cytogenetically.

*Bodmer:* First of all this isn't a tumour. Second, of the colorectal cell lines that we work with, while they all have some chromosomal changes, a number of them are 'near diploid'. It is much more significant to find a particular allele loss in a near diploid tumour. For example, we have one cell line where there are three recognizable cytogenetic events: an iso1p, an iso14p chromosome and an X;17 translocation. Then, apart from the monosomy for 17p, you can be reasonably sure that allele loss in other chromosomes reflects a specific genetic change. If, however, you have a highly heteroploid tumour, you can't be sure of the significance of allele loss.

*Knudson:* I agree with you about this polyposis. We and Philip Fialkow have used that same argument to explain the polyclonal nature of neurofibromas in von Recklinghausen's disease.

Loss of heterozygosity for 17p markers has also been reported in osteosarcomas, but not in retinoblastomas. It is well known that retinoblastomas are always near diploid, whereas tetraploid cells are frequently found in osteosarcomas. It is possible that some losses of heterozygosity could come from a tetraploid intermediate, with later reversion to diploidy, sometimes causing loss of heterozygosity on several chromosomes.

*Bodmer:* There is a problem in being sure of the significance of allele loss when you find too much of it. I, however, think it is unlikely that you go from tetraploidy back to diploidy.

*Benedict:* In studies done many years ago on *in situ* carcinoma of the cervix, it was shown that in the dysplastic lesions the chromosomal pattern was more often in the tetraploid range, whereas in the invasive carcinomas the chromosomal pattern was closer to the diploid number—although not diploid.

*Bodmer:* Again I would reverse your argument and say that I would want to be sure about really going back to diploidy versus near diploidy. We have one example: C. Paraskeva, when he was at the ICRF, grew a cell line from a polyp which is diploid with no detectable abnormality (Paraskeva et al 1984). It is a slowly growing cell line that seems to be normal in every respect and it is still heterozygous for chromosome 5 markers.

These changes are relatively specific—you do not find changes on lots of other chromosomes at the same time. The chromosome 22 change has not been confirmed in all cases; the chromosome 17 changes are very specific for 17p and there are two cell lines that are clearly monosomic for 17p. There is some variation in the data on the frequency of 17p loss. The frequencies may, however, vary according to the tumour type and the way the material for study has been chosen.

*zur Hausen:* I don't think the cervical dysplasias are good examples of polyploidy at an early stage preceding the development of a more aneuploid state of the precancerous lesion. Polyploid dysplasia lesions are clearly more frequently linked to specific types of viral infections rather than those which are near diploid or aneuploid.

*Benedict:* The work that I mentioned was by Boddington and Spriggs. They studied a series of dysplasias and cervical carcinomas *in situ* . They found that the dysplastic lesions were closer to the tetraploid range and that the carcinomas were not diploid but were closer to the diploid number.

*zur Hausen:* This effect is true, but in this case what is regarded as an early lesion is a lesion which is linked to specific types of viral infection, preferentially HPV6 and 11, as compared with those which show an early aneuploid state or near diploid state, such as HPV16 and 18 and others.

*Harris:* In *in vitro* systems it is very common to generate aneuploidy via tetraploidization. V. Defendi showed many years ago that when cells are infected with SV40 most of them become tetraploid after the first couple of divisions. Then they eliminate chromosomes at a great rate. That is a very common mechanism of karyotypic evolution.

*Bodmer:* But in cases like that the chromosome number doesn't fall to anything like diploid.

*Harris:* No, not strictly diploid, but in some of the cases that have been reported, the chromosome numbers dropped to something close to diploid.

*Benedict:* Does going from heterozygosity to homozygosity always mean loss of a genetic allele? In some of these situations could we be looking at unmasking of a truly recessive gene?

*Stanbridge:* That is very difficult to answer unless you have the gene available in a cloned form.

*Bodmer:* In the cases where there are deletions, like in retinoblastoma and familial polyposis, there is a correlation between the deletion being inherited within a family and the phenotype. To say that it is total loss of function is not necessary; by the models I've suggested one could envisage a recessive polyposis where the level of gene product is reduced by 50% and you see the effect only in the homozygote; nevertheless, that could be an early event which allows other genetic changes in a chromosome 5 to take place.

## References

Bodmer WF, Bailey CJ, Bodmer J et al 1987 Localization of the gene for familial adenomatous polyposis on chromosome 5. Nature (Lond) 328:614–616

Bos JL, Fearon ER, Hamilton SR et al 1987 Prevalence of *ras* gene mutations in human colorectal cancers. Nature (Lond) 327:293–297

Fearon ER, Hamilton SR, Vogelstein B 1987 Clonal analysis of human colorectal tumors. Science (Wash DC) 238:193–196

Herrera L, Kakati S, Gibas I, Pietrzak E, Sandberg AA 1986 Gardner syndrome in a man with an interstitial deletion of 5q. Am J Med Genet 25:473–476

Monpezat J-Ph, Delattre O, Bernard A et al 1988 Loss of alleles on chromosome 18 and on the short arm of chromosome 17 in polyploid colorectal carcinomas. Int J Cancer 41:404–408

Okamoto M, Sasaki M, Sugio K et al 1988 Loss of constitutional heterozygosity in colon carcinoma from patients with familial polyposis coli. Nature (Lond) 331:273–277

Paraskeva C, Buckle BG, Sheer D, Wigley CB 1984 The isolation and characterization of colorectal epithelial cell lines at different stages in malignant transformation from familial polyposis coli patients. Int J Cancer 34:49–56

Solomon E, Voss R, Hall V et al 1987 Chromosome 5 allele loss in human colorectal carcinomas. Nature (Lond) 328:616–619

Vogelstein B, Fearon ER, Hamilton SR et al 1988 Genetic alterations during colorectal-tumor development. New Engl J Med 319:525–532

# The molecular basis of retinoblastomas

Robert A. Weinberg

*Department of Biology, Massachusetts Institute of Technology, Whitehead Institute of Biomedical Research, Nine Cambridge Center, Cambridge, Massachusetts 02142, USA*

*Abstract.* Retinoblastomas and osteosarcomas in humans are associated with the homozygous inactivation of the *Rb* gene which maps to a locus on chromosome 13q14. This gene has been isolated as molecular clones. The gene encodes a protein of 105 kDa which is a nuclear DNA-binding phosphoprotein. The protein is not found in 15 retinoblastomas studied so far. An independent line of work examining the transforming mechanisms of the adenovirus E1A oncogene has shown that its encoded oncoprotein complexes with the *Rb*-encoded p105 in virus-transformed cells. This suggests that the *Rb* protein acts at a central intersection in the cell's growth regulatory network. Its alteration by genetic or epigenetic mechanisms appears to represent a critical event in several different pathways of tumorigenesis.

*1989 Genetic analysis of tumour suppression. Wiley, Chichester (Ciba Foundation Symposium 142) p 99–111*

Over the past decade, research into cellular oncogenes has provided great insight into genes and gene products that regulate normal and neoplastic growth (Bishop 1983, Varmus 1984). While this large and diverse field is difficult to summarize, several common and unifying threads run through much of the work that has been reported. The first of these is that oncogenes all act as agonists of cell growth; they appear to be hyperactive alleles of normal cellular growth-promoting genes. Consequently, oncogenic alleles act dominantly in relationship to the normal proto-oncogene alleles from which they arise.

A second common feature concerns the mode of activation of these dominant alleles: somatic genetic changes are invariably responsible for creating oncogenes. These somatic mutations may occur because of random errors of replication or the actions of carcinogens, but they have not yet been observed in germ cells. Accordingly, oncogenes cannot be enlisted to explain or understand the genetic elements that are transmitted in the germline and confer increased susceptibility to cancer.

A distinct class of genes, referred to here as tumour suppressor genes or anti-oncogenes, has been discovered in the past years. Its members act in ways that are diametrically opposed to the mode of action of oncogenes (Klein 1987). Thus, it appears that genes of this type act to constrain or antagonize growth in the normal cell. Moreover, cancer-promoting alleles of these genes can be passed through the germline and serve as determinants of inherited cancer susceptibility.

The best-studied of these genes is that which predisposes to retinoblastomas and sarcomas. The DNA sequence corresponding to this locus has been isolated and characterization of this sequence has begun to yield insight into the mode of action of the *Rb* gene-encoded protein.

## Genetics of retinoblastoma

The disease of retinoblastoma is currently observed in two manifestations — sporadic and familial. The sporadic cases are seen in young children in which no preexisting familial predisposition has been observed. In the familial disease, a similarly afflicted parent can be ascertained, half of whose children generally contract this otherwise very rare disease. Knudson (1971) argued that both forms of the disease reflect the presence of two mutant alleles in a tumour cell genome. He suggested that in sporadic cases of retinoblastoma, both required mutational events are sustained somatically in one or another retinal stem cell, whereas in the contrasting familial cases, one of the essential mutations is already present in the conceptus and thus is carried in all descendant cells, including all cells of the retina. The second required mutation is then sustained somatically in one of these already singly mutated cells.

This hypothesis has since been vindicated. A centrally important result came from the work of Yunis, who found interstitial deletions involving chromosome 13 band q14 in a number of retinoblastomas (Yunis & Ramsay 1978). The repeated linkage of such karyotypic abnormalities with the tumour phenotype supported the conclusion that this band carries at least one gene whose genetic inactivation underlies tumorigenesis. Such inactivation is the antithesis of the mutational events that activate proto-oncogenes and convert them to oncogenes.

Subsequent studies using a closely linked marker specifying the enzyme esterase D indicated that the second genetic target in the cell genome is the surviving, wild-type, homologous copy of the *Rb* gene (Godbout et al 1983). In many tumours, this hitherto intact allele is replaced by a copy of the already mutated allele, this being reflected by not only the homozygosity of the *Rb* gene but that of neighbouring loci as well.

The end result of these genetic changes is the full loss of *Rb* function. Knudson's initial hypothesis can now be understood in terms of these homozygous inactivations of function (see Knudson, this volume). An inescapable conclusion is that the *Rb* gene functions as a growth antagonist within the normal cell. The heritable cancer-inducing alleles in humans must therefore be null alleles that are phenotypically silent in an afflicted child until a complementing somatic mutation removes its surviving wild-type homologue in an appropriate target cell.

**Cloning of the Rb gene**

The isolation of DNA sequences having properties of the *Rb* locus depended on the creation of a library of chromosome 13-specific lambdaphages (Lalande et al 1984). From this a series of probes was generated and used to survey the DNAs of more than 50 retinoblastomas. One such probe, termed H3-8, was found to recognize deletions in as many as 30% of retinoblastomas (Dryja et al 1986, Friend et al 1986). Further work established that this probe derives from the middle of a gene having all the properties of the *Rb* gene itself (Friend et al 1986). The gene encompasses as much as 200 kbp of DNA. It is invaded by deletions from either rightward- or leftward-lying chromosomal regions in various tumours; occasionally, deletions affecting only internal parts of the gene are detected. These results have been substantially extended and reinforced by the work of two other groups which have cloned the gene through use of the H3-8 probe (Lee et al 1987a, Fung et al 1987). In summary, this gene appears to be the repeated target of the random genetic events that trigger retinoblastomas.

Several ostensibly discordant results would seem to cloud the above conclusion. The first stems from attempts to detect lesions in the candidate *Rb* gene by use of the Southern blotting procedure. Of the large number of tumours examined, only 10% yield lesions affecting both alleles, while 20% yield results indicating that only one of the two gene copies has been affected by deletion (Friend et al 1986, Fung et al 1987). This seems to conflict with the prevailing presumption that inactivation of the *Rb* gene underlies the genesis of virtually all retinoblastomas. We suggest, however, that Southern blotting is only a relatively crude procedure with which to detect inactivating mutations of the *Rb* gene. The probe used was a cDNA probe and as such will only sense changes that affect exons. Moreover, relatively minor changes in gene structure, which may be undetectable by blot analysis, are perfectly capable of inactivating gene function.

Recently, we and collaborators have detected such a subtle lesion in the genome of a bladder carcinoma, a tumour in which *Rb* inactivation has not been postulated to play a major role. Work in the laboratory of Ed Harlow of Cold Spring Harbor indicated the absence of a functional *Rb* gene product in cells of the J82 bladder carcinoma cell line by tests that I describe below (E. Harlow, unpublished work 1988). Analysis in my laboratory by Jonathan Horowitz showed that a gene product was indeed present but apparently lacking about 3 kDa of protein sequence, as estimated from electrophoretic migration rates (J. Horowitz, unpublished work 1988). Sang Ho Park then found, using a polymerase chain reaction, that a 105 bp internal segment was missing from the reading frame of the cDNA encoded by the *Rb* gene, leaving an in-frame deletion that would account for the observed truncation of the gene product (S.H. Park, unpublished work 1988).

Subsequent work by David Yandell and Thaddeus P. Dryja of the

Massachusetts Eye and Ear Infirmary showed that the deleted segment corresponds precisely with an exon of the gene and that this exon is missing because of a point mutation in a splice acceptor site which is required for the proper processing of the mRNA precursor and for the inclusion of this exon in the final processed mRNA (D. Yandell, T. Dryja, unpublished work 1988). This suggests that numerous subtle mutations of the *Rb* locus will be found in future years that will prove as effective as major deletions in knocking out gene function while eluding detection by all but the most detailed examinations of allele structure.

A second set of observations also seems to disagree with the identification of the cloned DNA with the *Rb* locus. This concerns the tissue specificity of tumour induction associated with *Rb* inactivation and the contrasting range of tissues in which the candidate gene is expressed. When defining the tissue specificity, it is important to note that *Rb* inactivation is not associated exclusively with the triggering of retinoblastomas. Indeed, children congenitally afflicted with a defective *Rb* allele run greatly increased risks of osteosarcomas (Draper et al 1986). Accordingly, the term '*Rb*' is a misnomer in that these connective tissue tumours are also predisposed by defects in this gene. Nonetheless, retinoblastomas and osteosarcomas do represent a distinct subset of the tissues and tumour types encountered in oncology.

Contrasting with this narrow range of associated tumour types, we note that transcripts reactive with the cloned gene in Northern blot analysis are expressed in a wide range of tumours and normal tissues. Greg Shackleford and Harold Varmus in San Francisco have detected these 4.7 kb transcripts in a wide range of tissues in the developing mouse, using as probe a mouse cDNA homologous to the cloned human gene and isolated by Rene Bernards (R. Bernards & R.A. Weinberg, unpublished work 1988). Thus, the cloned gene is expressed in a wide range of tissues and apparently plays a role in the growth regulation of these tissues, yet its inactivation triggers neoplasia in only a narrow subset of cell types. This paradox is not resolved by presently available data. It may be years before we understand why *Rb* inactivation affects so few cell types in the body.

## The *Rb* gene product

As described above, the cloned gene specifies a 4.7 kb transcript from which a cDNA has been prepared by several groups. Sequence analysis of the cDNA by the group of Wen-Hwa Lee and by Stephen Friend of my own group predicts the synthesis of a 105–110 kDa protein (Lee et al 1987b, Friend et al 1986). Lee's group has detected such a protein in the nuclear fraction of certain cells, a result confirmed by my own group (J. Horowitz, unpublished work 1988). In both instances, synthetic oligopeptides whose sequences were derived from translation of the cDNA reading frame were used to elicit the production of antisera that reacted with this protein.

The p105 encoded by the gene is a phosphoprotein that has affinity for DNA cellulose (Lee et al 1987b). This fact, taken together with other properties of the protein (see below), raises the possibility that this protein functions as part of the nuclear apparatus directly regulating gene expression. Such a finding would provide a counterpoint to work on nuclear oncogene proteins, such as those encoded by the *fos* and *jun* genes, which have been shown to be directly involved in DNA binding and in associated regulation of transcription (Lech et al 1988, see Struhl 1988). By use of these antisera, Horowitz has found that the p105 is absent from 15 retinoblastoma tumours examined to date (J. Horowitz, unpublished work 1988). This further supports the idea that this protein is the gene product of the *Rb* locus.

## Interaction between *Rb* and a viral oncogene

An apparently unrelated line of research has recently converged with work on the *Rb* gene in a striking and totally unexpected way. This research began in the laboratories of Phillip Branton and Ed Harlow, both of whom were interested in the mechanism of action of the E1A oncogene of human adenovirus. This oncogene is known for its multifarious activities: it is involved in immortalization of cultured cells, it collaborates with a *ras* oncogene in conferring a tumorigenic phenotype, and it acts as a *trans*-activator or repressor of a series of viral and cellular genes (Berk 1986).

The two research groups investigated these functions by ascertaining whether the E1A oncoproteins bound any proteins of the host cell. Using anti-E1A monoclonal antibodies, they found that the E1A proteins are complexed with as many as 8–10 distinct host cell proteins in virus-transformed cells. These host proteins presumably represent targets that are acted upon by the viral oncoprotein (Yee & Branton 1985, Harlow et al 1986). Changes in their function following complex formation may be responsible for some of the effects induced by the E1A gene.

One of the complexed proteins is a phosphoprotein of 105 kDa whose properties were seen to resemble those associated with the *Rb* protein. Peter Whyte of Harlow's group went on to show that the E1A oncoprotein was able to complex with this p105 protein in a variety of tumour types but not in retinoblastomas. It remained only to demonstrate an identity between the E1A-associated p105 and the *Rb*-encoded p105. This was achieved by collaboration between Harlow's group and Jon Horowitz working in my own laboratory (Whyte et al 1988).

This finding has wide conceptual ramifications. It shows a direct confrontation between a viral oncogene and a cellular 'anti-oncogene' via a physical interaction of their two proteins. If the absence of the *Rb* gene product leads to cell transformation, then the oncogene may achieve the same physiological result through the complexing of its gene product with the *Rb* protein. By such logic, the E1A protein exerts an inhibitory effect on the functioning of

*Rb*-p105 proteins with which it is complexed. Since E1A is a transcription regulator, it is plausible that *Rb*-p105 also is involved in modulating the transcription of critical target genes.

## The *Rb* gene and cell physiology

How can *Rb* function be rationalized in terms of the physiology of the normal cell? We favour the theory that a normal stem cell is equipped to respond to both growth and antiproliferative signals. Moreover, we speculate that the *Rb* protein is a component of the machinery that enables the stem cell to respond to antiproliferative signals in its environment. Loss of the *Rb* gene product would then cause loss of responsiveness to such signals without necessarily affecting the ability of such cells to respond to mitogens. Such antiproliferative signals might include those that induce differentiation, leading to end-stage cells that cease growing. The end result of this deficit in a signalling pathway would be the loss of a critical regulatory mechanism that is essential for restraining stem cell proliferation.

One example of such loss of responsiveness arises from studies that we undertook with Adi Kimchi of the Weizmann Institute (Kimchi et al 1988). These showed that retinoblastoma cells, in contrast to normal human fetal retinal cells, do not respond to TGF-β. While the growth of normal retinal cells is strongly inhibited by TGF-β, that of retinoblastoma cells is unaffected. This was found to be due to the loss of TGF-β receptors. Unlike most normal and most malignant cell types, which display as many as three distinct forms of TGF-β receptor, retinoblastoma cells have no normal receptors. We conclude that loss of the receptors, which may be secondary to loss of the *Rb* gene product, allows retinoblasts to escape the inhibitory effects of TGF-β, a factor normally present in the developing retina. Thus, one of the growth regulatory pathways affected by the *Rb* gene seems to be revealed, indicating one of the molecular mechanisms through which *Rb* loss leads to malignancy.

Over the coming years, studies at the levels of biochemistry and cell physiology will address this model and the way in which signal transduction pathways are organized to limit cell proliferation.

*Acknowledgements*

This work was supported by grants from the NIH, American Business Foundation for Cancer Research and the Rowland Foundation. R.A.W. is an American Cancer Society Research Professor.

## References

Berk AJ 1986 Adenovirus promoters and Ela transactivations. Annu Rev Genet 20:45–79

Bishop JM 1983 Cellular oncogenes and retroviruses. Annu Rev Biochem 52:301–354
Draper GJ, Sanders BM, Kingston JE 1986 Second primary neoplasms in patients with retinoblastomas. Br J Cancer 53:661–671
Dryja TP, Rapaport JM, Joyce JM, Petersen RA 1986 Molecular detection of deletions involving band q14 of chromosome 13 in retinoblastomas. Proc Natl Acad Sci USA 83:7391–7394
Friend SH, Bernards R, Rogelj S, Weinberg RA, Rapaport J, Albert D, Dryja TP 1986 A human DNA segment with properties of the gene that predisposes to retinoblastoma and osteosarcoma. Nature (Lond) 323:643–646
Friend SH, Horowitz JM, Gerber MR, Wang XF, Bogenmann E, Li FD, Weinberg RA 1987 Deletions of a DNA sequence in both retinoblastomas and mesenchymal tumors: organization of the sequence and its encoded protein. Proc Natl Acad Sci USA 84:9059–9063
Fung Y-KT, Murphree AL, T'Ang A, Qian J, Hinrichs SH, Benedict WF 1987 Structural evidence for the authenticity of the human retinoblastoma gene. Science (Wash DC) 236:1657–1661
Godbout R, Dryja TP, Squire J, Gallie BL, Phillips RA 1983 Somatic inactivation of genes on chromosome 13 is a common event in retinoblastoma. Nature (Lond) 304:451
Harlow E, Whyte P, Franza BR, Schley C 1986 Association of adenovirus early-region 1A proteins with cellular polypeptides. Mol Cell Biol 6:1579
Kimchi A, Wang X-F, Weinberg RA, Cheifetz S, Massagué J 1988 Absence of TGF-β receptors and growth inhibitory responses in retinoblastoma cells. Science (Wash DC) 240:196–199
Klein G 1987 The approaching era of the tumor suppressor genes. Science (Wash DC) 238:1539–1545
Knudson AG 1971 Mutation and cancer: statistical study of retinoblastoma. Proc Natl Acad Sci USA 68:820–823
Knudson AG 1989 Epidemiology of genetically determined cancer. In: Genetic analysis of tumour suppression. Wiley, Chichester (Ciba Found Symp 142) p 3–19
Lalande M, Dryja TP, Schreck RR, Shipley J, Flint A, Latt SA 1984 Isolation of human chromosome 13-specific DNA sequences from flow sorted chromosomes and potentially linked to the retinoblastoma locus. Cancer Genet Cytogenet 13:283–295
Lech K, Anderson K, Brent R 1988 DNA-bound fos proteins activate transcription in yeast. Cell 52:179–184
Lee WH, Bookstein R, Hong F, Young L-J, Shew J-Y, Lee EY-HP 1987a Human retinoblastoma susceptibility gene: cloning, identification and sequence. Science (Wash DC) 235:1394–1399
Lee H-W, Shew J-Y, Hong FD et al 1987b The retinoblastoma susceptibility gene encodes a nuclear phosphoprotein associated with DNA binding activity. Nature (Lond) 329:642–645
Struhl K 1988 The JUN oncoprotein, a vertebrate transcription factor, activates transcription in yeast. Nature (Lond) 332:649–650
Varmus HE 1984 The molecular genetics of cellular oncogenes. Annu Rev Genet 18:553–612
Whyte P, Buchkovich KJ, Horowitz JM et al 1988. Association between an oncogene and an anti-oncogene: the adenovirus E1A proteins bind to the retinoblastoma gene product. Nature (Lond) 334:124–129
Yee S-P, Branton PE 1985 Detection of cellular proteins associated with human adenovirus type 5 early region 1A polypeptides. Virology 147:142
Yunis JJ, Ramsay N 1978 Retinoblastoma and subband deletion of chromosome 13. Am J Dis Child 132:161

## DISCUSSION

*Bodmer:* It's still an interesting paradox why there is not a broader range of tumours in the families with retinoblastoma. There, presumably, the gene is deleted and you might expect to see a range of tumours that matches to some extent the tissues in which you see somatic cell changes.

*Weinberg:* I agree and I cannot provide a good explanation. As Wen-Hwa Lee and co-workers have observed, 10–20% of mammary carcinomas have changes in the *Rb* gene arising by somatic mutations. Why, therefore, doesn't one see an increased incidence of mammary carcinomas in families with retinoblastoma?

*Adams:* Perhaps the restricted range of tumours associated with *Rb* deletion means that in many cell types the *Rb* gene product is essential for cell viability. If so, its loss would simply eliminate that clone rather than generate a tumour.

*Weinberg:* One idea is that in many cell types there are several redundant negative growth regulatory circuits, acting in parallel, in which case the loss of one such circuit will not lead to uncontrolled growth and a tumour. But in certain lineages there may be only one such gene actively expressed, for example, in the cone lineage of photoreceptors and many mesenchymal lineages. In these particular cell types, the *Rb* gene may represent the only barrier to uncontrolled proliferation.

*Harris:* The problem is that, as far as I know, adenovirus is not involved in any human tumour, and SV40 is pretty well in the same category. The only one of the viruses you mention that looks relevant to human tumours is papillomavirus, but this is not involved in retinal tumours. It's not generalized models of tumorigenicity that we need, but an explanation of why retinoblastomas arise. Until one begins to talk about mechanisms that regulate the differentiation of a retinal cell, one is not going to provide a satisfactory explanation of the genesis of retinoblastoma.

*Weinberg:* I will make the following suggestion—retinoblastomas have provided us with a window on a very interesting gene, the *Rb* gene. I suggest that the *Rb* gene acts as a growth regulator in a wide variety of tissues, not just retinal cells. For example, in mesenchymal cells a defect in the *Rb* gene can lead to sarcomas. It may be a critical growth regulator in a variety of other tissues as well, and in those tissues its deletion may also provoke tumour development. It is not involved in human tumours of viral aetiology, with the exception of human cervical carcinoma, in which human papillomavirus is strongly implicated. Here, the E7 oncoprotein of this virus seems to complex with the *Rb* protein, according to the work of Peter Howley and Ed Harlow.

The fact that three different, ostensibly independently evolved, DNA tumour viruses each have an oncoprotein which complexes with the *Rb* gene protein suggests to me that this protein is an important regulator of cell growth in many different cell types. We can assume that rodent models of tumorigenicity and carcinogenesis provide very good models at the molecular level for what's happening in humans. I assume therefore that these data will ultimately be extended to human tumour models where loss or inactivation of the *Rb* gene or its product will be found to play a causal role in a number of different tumours.

*Sporn:* Many growth factors exist in alternative and multiple forms, although some tissues may not have all of those. You could postulate that in the retina some of the redundancy that may exist in other epithelial tissues is not present; a loss in that case will give rise to a tumour. In some of the other epithelial tissues the additional redundancy compensates for that loss. Once the molecules have been identified, we will be able to look for these isoforms and may find that the isoforms don't exist in the retina and they do in other tissues.

*Varmus:* In the adenovirus-transformed retinal cell, the level of TGF-β receptor is normal. It is apparent that the complex formed between E1A and the *Rb* protein in that cell doesn't detectably alter the level of TGF-β receptor. If you argue that reduction of TGF-β receptor level in retinoblastomas is involved in tumorigenesis, then adenovirus is not working in the same way through the *Rb* protein.

*Weinberg:* There are two ways to respond to that. One is that the reduction in the amount of TGF-β receptor may not be causally related to loss of the *Rb* gene. They could represent two independent events, although I don't believe they do. I find it hard to conceive that loss of the TGF-β receptor occurs as a result of independent somatic mutations having nothing to do with the loss of *Rb* gene product. Secondly, we don't really understand the effects of adenovirus E1A binding to *Rb* protein, it may not mimic the loss of *Rb* protein that occurs when the gene is lost.

*Sporn:* In your paper with Adi Kimchi (1988), as a control experiment you found that these cells are making almost no TGF-β. In several other systems, it has been shown that growth factors are often involved in the induction of their own receptor. Most tumour cells make appreciable amounts of TGF-β in culture.

*Weinberg:* The reason we checked whether or not these tumour cells were secreting TGF-β is as follows: had they been secreting TGF-β, then the absence of receptor might have been attributable to down-regulation of receptor after ligand binding.

*Stanbridge:* Do retinoblastomas lack TGF-β?

*Weinberg:* We only looked at a few, but, of those, none was secreting TGF-β.

*Lau:* Bob, in order to address the question of whether the absence of TGF-β receptor is related to the loss of the *Rb* gene, could you microinject p105 into the retinoblastoma cells and look for re-expression of the TGF-β receptor?

*Weinberg:* In principle one could do that, but at the moment we cannot get sufficient amounts of the protein expressed. Bill Benedict has done a formally similar experiment with an osteosarcoma cell that had lost the TGF-β receptor and reacquired it when chromosome 13 was restored.

*Benedict:* Eric (Stanbridge) is going to discuss the addition of chromosome 13 to the osteosarcoma cell line, OSV, which has a truncated message for the *Rb* gene and we think does not contain any *Rb* protein. There are also no TGF-β receptors in these OSV cells, as shown by studies done in collaboration with Brian Carr. When the normal chromosome 13 was transferred into the OSV cells by microcell fusion, there was an inhibition of tumorigenicity in nude mice and at that time TGF-β receptor was expressed in these cells.

*Weinberg:* That's a very exiciting result because it indicates that there is some causal connection between the expression of *Rb* gene product and expression of the TGF-β receptor genes.

*Benedict:* There may be. We have found other tumour cells that lack the TGF-β receptor: HeLa cells and the retinoblastoma cell line, Y-79. When Dr Steve Pasquale in Dr Bernard Weissman's laboratory fused Y-79 with HeLa, a suppressed cell hybrid was produced in which TGF-β receptor was expressed. From that suppressed hybrid we obtained a tumorigenic revertant that seems to be missing the normal mRNA for the *Rb* gene. When chromosome 13 was put into that revertant hybrid, there was again suppression of tumorigenicity. We have not yet looked at the TGF-β receptor in these cells.

*Wyke:* I am now a little confused—is it general for the receptor to be missing in osteosarcomas and soft tissue sarcomas that lack the *Rb* gene?

*Weinberg:* It is generally thought that in retinoblastomas the receptor is missing. We looked at one osteosarcoma and found the TGF-β receptor. I don't know whether anyone has done a survey of the TGF-β receptor in osteosarcomas.

*van der Eb:* Is the *Rb* gene expressed in adult retina cells?

*Weinberg:* Yes, it's expressed at very high levels.

*van der Eb:* I would like to describe some preliminary experiments that we have been doing recently. They concern transformation by human adenoviruses, and also some data on SV40 and BK virus.

The transforming region of the adenoviruses is located in the 4000 bases at the left hand end of the viral genome, which correspond to early region I. This region, E1, can be subdivided into two transcription units, E1A and E1B. E1A codes for two mRNAs and EB1 for one mRNA. The two mRNAs of E1A encode two virtually identical proteins. E1B mRNA is translated into two unrelated proteins.

Very little is known about how E1A transforms cells. E1A can immortalize cells, it is important in transformation and it can suppress as well as activate the expression of other genes. The suppressing activity is associated with conserved region I (CRI) and CRII, which are conserved among the human adenoviruses. The activating function is associated with CRIII. A logical deduction might be that adenovirus transformation is the result of alterations in cellular gene expression.

To test this idea we investigated the expression of a number of genes in adenovirus-transformed cells and found that a number of serum or TPA-inducible genes are inhibited. We looked at the *JE* gene isolated by Cochran et al (1983), c-*myc*, which presumably has a role in regulating cell proliferation, and two genes coding for excreted proteases, collagenase I and transin or stromelysin. Expression of all these genes is suppressed in adenovirus-transformed cells. This suppression is a function of region E1A and occurs at the level of transcription, which indicates that the promoter is involved. We also know from using E1A mutants that for suppression of the *JE* gene, and probably also of the collagenase gene, CRI is responsible and probably not CRII. We have some evidence that the AP-1 binding site which is present in JE, collagenase and transin is involved in this inhibition, but there is no formal proof for that yet. We have also observed recently that glucocorticoid hormones, such as dexamethasone, have a similar effect on expression of these genes as does E1A, suggesting that E1A might utilize a similar inhibitory mechanism as these hormones.

We have noticed that E1A can transform rodent cells quite efficiently, but for some reason it does not transform human cells. Van der Noordaa and his co-workers (De Ronde et al 1988) in Amsterdam studied the transformation of human diploid fibroblasts by BK virus. Whereas the SV40 early region can efficiently transform primary baby rat kidney cells and diploid human fibroblasts, the BK virus cannot transform the diploid human fibroblasts, although it efficiently transforms the rodent cells. Van der Noordaa had not been able to determine the reason for this behaviour. He then noticed that the SV40 early region in which the small T antigen encoding area had been partially deleted has the same phenotype as the BK virus, in that it can easily transform baby rat kidney cells but not human fibroblasts. Then, by chance, he tested diploid fibroblasts from a patient with Wilms' tumour in whom one of the chromosomes 11 contains a deletion. Both the BK virus and the small T mutant of SV40 can transform these cells, although at lower efficiency. This suggests that the small T antigen of SV40 may be able to partially inactivate the product of the gene that has been deleted in this Wilms' tumour fibroblast. The BK virus small T antigen does not have this ability.

We then tested whether the diploid fibroblasts from this Wilms' tumour patient could be transformed by adenovirus: they could not. We therefore looked at other cells from individuals prone to inherited cancer. The following cells were tested: normal diploid skin fibroblasts, diploid fibroblasts from patients with Wilms' tumour, diploid skin fibroblasts from patients with retinoblastoma, and diploid human embryonic lung cells, embryonic retinoblasts and embryonic kidney cells. Wild-type SV40 transforms all these cells. The SV40 small T antigen-deficient mutant does not transform the normal cells, but it does transform the diploid Wilms' tumour fibroblasts; it does not transform human embryonic lung but, surprisingly, it does transform normal human embryonic retinoblasts. Adenovirus E1 does not transform any of these cells

(Gallimore et al 1986, A G Jochemsen, unpublished work), except the human embryonic retinoblasts. Human embyronic kidney cells are transformed only rarely and we observed only two transformed clones in a large number of experiments in these cells.

To investigate whether the SV40 small T antigen mutant and adenovirus E1 or E1A can complement each other in transformation of normal diploid human fibroblasts, we co-transfected these and some of the other cells types with the SV40 small T mutant and adenovirus region E1A. No transformation was seen with the normal diploid skin fibroblasts; a low SV40-like transforming activity was seen with Wilms' tumour fibroblasts (indistinguishable from that obtained with the SV40 small T antigen mutant alone). Surprisingly, there was a high transforming activity in the human embryonic retinoblasts. It is as if E1A complements this small T defect, because the transformation is like that induced by SV40. This was unexpected because E1A is known to suppress the SV40 enhancer. We are going to try more diploid fibroblasts from individuals genetically predisposed to cancer.

*Benedict:* Dr van der Eb, could you define what you mean by transformation?

*van der Eb:* Transformation in this case is morphological transformation in a focus assay. BK virus can be expressed in normal human fibroblasts without the cells changing their morphology. This is information from Van den Noordaa's laboratory. However, the Wilms' tumour fibroblasts are morphologically transformed in a focus assay. So this does not concern tumour suppression, but suppression of transformation.

*Benedict:* So the diploid Wilms' fibroblasts have not been immortalized by the virus?

*van der Eb:* No, I predict that none of these cells are immortalized.

*Benedict:* Dr Banerjee in our laboratory has transfected fibroblasts from patients with retinoblastoma using constructs that contain SV40. They have all given rise to immortalized cells, although so far none of them has been tumorigenic.

*van der Eb:* How sure are you that they are truly immortalized?

*Benedict:* They have been in culture for more than two years, they are at passage 90 or 100, they are aneuploid, grow in semi-solid medium and they have all the characteristics of transformed cells.

*van der Eb:* Diploid human fibroblasts transformed by SV40 can have an enormously extended lifespan, they can go up to 120 passages and then enter crisis.

*Benedict:* These cells have all been through crisis.

*Varmus:* Work from Ed Ziff's lab has dissociated the transcriptional suppression of E1A from its transforming activity. Have you looked in human fibroblasts to ask whether adenovirus E1A continues to suppress TPA-inducible genes?

*van der Eb:* We have not tested that particularly, except for human embryonic retinoblasts, in which the collagenase gene is suppressed.

## References

Cochran BH, Reffel AC, Stiles CD 1983 Molecular cloning of gene sequences regulated by platelet derived growth factor. Cell 33:939–947

Gallimore PH, Grand RJA, Byrd PJ 1986 Transformation of human embryo retinoblasts with simian virus 40, adenovirus and RAS oncogenes. Anticancer Res 6:499–508

De Ronde A, Mannens M, Slater RM et al 1988 Morphological transformation by early region human polyomavirus BK DNA of human fibroblasts with deletions in the short arm of one chromosome 11. J Gen Virol 69:467–471

Kimchi A, Wang X-F, Weinberg RA, Cheifetz S, Massagué J 1988 Absence of TGF-β receptors and growth inhibitory responses in retinoblastoma cells. Science (Wash DC) 240:196–199

# General discussion III

## Transforming growth factor-β: a negative regulator

*Sporn:* First I should acknowledge Anita Roberts, who has been my collaborator in all of our work that I will discuss; her contributions have been most important. As a general concept one can state that any loss in a negative regulatory system involved in growth control could lead to a malignant phenotype. TGF-β happens to be one of the first peptides to be identified which has this growth regulatory effect. We can study this process by investigating the loss of TGF-β itself, or its receptor, or any element in its signalling system. Even the loss of a unique nuclear transcription factor which might be required for the expression of the gene for TGF-β or its receptor would be significant. One might think of proteins like the *Rb* protein in this context.

TGF-β is a disulphide-linked homodimer, 25 kDa molecular weight. Its sequence is highly conserved across many species and it has been isolated from many different tissues. It is secreted in a biologically latent form and almost all cells have TGF-β receptors of very high affinity. I think the recent examples of tumour cells that have no receptors are the exception. There are at least four functionally similar and highly homologous proteins in this family: β1 was the first to be discovered; β2 is well characterized; β3 has been cloned recently in three laboratories, including our own, and Sonia Jakowlew in our lab has just cloned chicken β4.

These proteins have several roles, but I will discuss only the anti-proliferative one. In bronchial epithelial cells there is a very high degree of sensitivity—the $ED_{50}$ for inhibition of proliferation of many types of epithelial cell is in the order of only a few picograms/ml. The growth inhibition can be seen *in vivo* as well as *in vitro*. *In vivo* implantation of TGF-β in the mouse mammary gland markedly inhibits epithelial growth. This is Charles Daniel's and Gary Silberstein's work (1987). It is the opposite of the proliferative results that Harold Varmus described with *int-1*. In many situations this anti-proliferative effect of TGF-β possibly overrides the effects of other positive growth factors, such as epidermal growth factor (EGF), fibroblast growth factor (FGF) or interleukin-2. Growth inhibitory effects of TGF-β are seen in many immune cells, T cells and B cells in particular, and again the growth inhibition is very striking. Human TGF-β2 was first purified by its immunosuppressive effects. It is a product of human glioblastoma cells and was purified using an assay which measured inhibition of thymocyte proliferation.

Several years ago we suggested that transformation might be associated with a variety of mechanisms involving the loss of negative autocrine activity: these include mutation or deletion of the gene for a negative effector itself, failure to process such an effector properly, or the loss of receptors for such an effector. Recently, Andrew Geiser has been working with us on some fusion hybrids that were made in Eric Stanbridge's lab. These are fusions of the EJ bladder carcinoma cell and SW480 cell. Both of the parental cell lines, which are highly malignant, have completely lost their sensitivity for TGF-β, and continue to grow well in its presence. Why EJ has lost this sensitivity we don't know; SW480 is probably one of these rare cells that does not have the receptor, although that is somewhat controversial. Interestingly, the hybrids are non-tumorigenic and they have regained sensitivity to growth inhibition by TGF-β. There is also a tumorigenic hybrid and that has retained its insensitivity.

The other protein which we are very excited about as a negative growth regulator is the retinoic acid receptor protein that's recently been cloned in Pierre Chambon's and Ronald Evans' laboratories. There are two retinoic acid receptors (RAR-α and RAR-β); they are both negative growth regulators. Defects in retinoic acid receptors that are associated with malignancy may be found. Retinoic acid can restore the non-malignant phenotype in fully malignant cells, such as HL60 promyelocytic leukaemia cells and F9 teratocarcinoma cells.

We are interested in whether there is any connection between the retinoid system and the TGF-β system. We asked the following questions: 1) do retinoids induce the expression of TGF-β itself? 2) do they induce the expression of the receptor for TGF-β? 3) do they modulate the TGF-β signalling system? Some recent work with Stuart Yuspa and Adam Glick shows that retinoic acid induces high levels of TGF-β secretion in primary cultures of mouse keratinocytes. Adam Glick and Seong-Jin Kim in my laboratory have cloned the promoter for TGF-β and in transfection studies we are starting to look at whether signals such as retinoids or steroids are involved in turning on TGF-β expression. I think TGF-β will prove to be clinically important in the suppression of the malignant phenotype. Ultimately, this could be most useful in the prevention of human cancer.

*Stanbridge:* This family is composed of at least four members, is there tissue-specific expression of any member of the family?

*Sporn:* We have been looking at that with TGF-β1 and β2. There are cells that make predominantly β1, for example human platelets, other cells make predominantly β2. Otherwise, not too much is known about tissue specificity and the β3 and β4 peptides have not been isolated yet. We only know those peptides at the gene level, although antibodies are being made from the deduced amino acid sequences.

*Sachs:* Are the other TGF-β genes very similar to β1 and 2? Is there much variation in the sequences?

*Sporn:* The peptides all have conservation of the cysteines in the critical positions which give this molecule its shape. For practical purposes, TGF-β1 and β2 are in almost all situations functionally interchangeable, although some significant differences will probably be discovered. We don't know about types 3 and 4, because we have only the DNA sequence, but those are entirely separate genes.

*Bodmer:* Do they map as a single cluster?

*Sporn:* We don't know chromosomal localizations for 3 and 4: I am not sure whether that's been done for TGF-β2.

*Adams:* Loss of the receptor for an inhibitory factor could be involved in the generation of neoplasia in many types of cells. For loss of the factor itself to be relevant, however, there would have to be autocrine production. Do many cells produce their own growth inhibitory factors?

*Sporn:* There are numerous examples in which that has been shown, although the action is often also paracrine.

*Wyke:* The complementary question is how good is TGF-β as a paracrine effector?

*Sporn:* That is particularly important in epithelial carcinogenesis, where you have to consider not only the epithelium but its relationship to the underlying stroma. The general paradigm is that when the epithelial cell becomes malignant it becomes insensitive to regulation by its own TGF-β, but at the same time it may be secreting a lot of TGF-β, which then stimulates angiogenesis and fibroblast proliferation in the underlying stroma. This gives the developing tumour a strong nutritive base. We have some histochemical results which support that general hypothesis.

*Balmain:* Epidermal cells or epithelial cells are an important source of TGF-β. Pre-malignant cells in papillomas are markedly inhibited by TGF-β, but cell lines from malignant carcinomas have lost that response. We have taken some immortalized cell lines which do respond, put activated *ras* genes into those, and then tested whether the *ras* gene switches off the response. The *ras* gene by itself does not switch off the response to TGF-β, as long as the cells remain non-tumorigenic. When that cell line becomes tumorigenic it loses the response to TGF-β. However, it is more complicated than that. There are other cell lines which spontaneously lose the response to TGF-β but are still non-tumorigenic, independent of *ras* expression. So there are two separate events which can play a rolė here.

*Adams:* Can you get profound effects on growth of any cell types with antibodies to TGF-β?

*Sporn:* The growth of some cells which are negatively regulated by TGF-β, for example the CCL64 lung epithelial cell line, is strongly stimulated by anti-TGF-β antibodies. This relates to some reservations that Bob Weinberg had a few years ago about this general autocrine hypothesis. These antibody studies provide strong evidence for a negative autocrine role of TGF-β.

*Harris:* Has anybody devised methods for measuring the concentration of any of these growth factors *in vivo*? Does one ever achieve *in vivo* the concentrations that are required to produce the *in vitro* effects? If you have antibodies, for example against TGF-β, you could measure, at least to a first approximation, whether you actually have enough TGF-β *in vivo*.

*Sporn:* You can measure serum levels; the level in serum is clearly growth inhibitory. However, with respect to a negative autocrine action on a cell, I think it would be extremely difficult to measure the concentration of factor present right at the surface of the cell. The plasma levels are sufficient to inhibit cell growth, but that may not be the actual functional growth inhibitor. The cell may be keeping itself in some growth suppressed state via the action of other inhibitors which may not diffuse very far from the cell.

*Harris:* The serum levels of the great majority of proteins are very different from the concentrations in the extracellular fluid. The question is whether any of these diagrams of autocrine circuits derived from *in vitro* studies are relevant to what happens *in vivo*.

*Balmain:* If you treat skin with the tumour promoter, TPA, then the amount of TGF-β messenger RNA increases. David Kerr at the Beatson Institute has shown that if you simply make extracts of the skin or the epidermis, you can detect biologically active TGF-β in those extracts at reasonably high levels. This correlates with some staining work done by Kathy Flanders in Michael Sporn's lab using anti-TGF-β antibodies, which showed the degree of staining increasing in conjunction with the biological activity.

*Marks:* For the haemopoietic growth factors the *in vivo* concentrations are in the range that produce an effect *in vitro*.

*Klein:* In our B lymphocyte series, TGF-β does not inhibit either lymphoblastoid cell lines or Burkitt's lymphoma cell lines, but it does inhibit the pro-B line I described in my paper (Klein, this volume). We cannot confirm Blomhoff's observation that EBV-negative Burkitt's lymphoma lines were inhibited but their EBV-converted derivatives were not.

*Sporn:* We have not done any work on that. This whole question of finding out where in the stages of progression or transformation the loss of sensitivity occurs and what is the mechanism is one of the key issues. Since we don't know what is involved in the signalling pathway between the receptor and the nucleus, it is a very difficult problem to look at.

*Nossal:* Active TGF-β is a profound inhibitor of antibody formation *in vitro* and is believed to act on both T cells and B cells. It also inhibits pre-B cell proliferation *in vitro*.

*Knudson:* Does stimulation by retinoic acid imply that the epithelial stem cell doesn't produce TGF-β?

*Sporn:* These keratinocytes, in culture at least, do not make much TGF-β. That agrees with Allan Balmain's report that there is very little TGF-β mRNA.

*Benedict:* We looked at some hybrids made in Dr Weissman's laboratory between tumorigenic cervical carcinoma HeLa cells and retinoblastoma Y79 cells. The hybrids were non-tumorigenic. We obtained a revertant cell that is now 100% tumorigenic. Chromosome 13 was put back into this tumorigenic segregant and tumorigenicity was again suppressed. With respect to the TGF-β receptor, neither Y79 nor HeLa cells have the receptor, whereas the suppressed Y79–HeLa hybrid does. This is similar to the results found with OSV cells, that is absence of TGF-β receptors in the OSV cells but reappearance of the receptors after introduction of a normal chromosome 13.

*Sporn:* Bill, when you put this chromosome 13 back into the OSV cells and the TGF-β receptor was again present, were those cells growth suppressed by TGF-β?

*Benedict:* Dr Hashimoto in our laboratory has done one quick study. Just looking at the cell numbers, over the first four days no differences in viability were seen after TGF-β was added to the suppressed cells.

*Marshall:* Dr Sporn, when you do these experiments of adding a growth factor and TGF-β together and you get inhibition of proliferation, do you know at what level that's operating? Do the early events still happen or where does it stop?

*Sporn:* We haven't done those experiments. Chambard & Pouysségur (1988) showed that all the early events are allowed to proceed. With EGF, platelet-derived growth factor and FGF, the original signalling events, i.e. tyrosine phosphorylation and kinase activity, are never blocked. The block occurs at an unknown, later stage of the cell cycle.

*Marshall:* So the first thing that you would observe would be inhibition of, for example, *myc* expression?

*Sporn:* The inhibition of *myc* has been seen by several workers. Early events never seem to be blocked, only the late events.

*Bodmer:* Aren't there probably at least two different pathways? TGF-β can increase the level of receptors for components of the extracellular matrix, which can then induce differentiation, thereby blocking growth. There are also direct effects that interfere with EGF- type signalling.

*Sporn:* There are a whole series of positive effects of TGF-β, which include turning on the extracellular matrix cascade, the expression of collagen, fibronectin and the fibronectin receptors. To the extent that any of these matrix substances are controlling cell proliferation by their own mechanism, then there is a two-tier regulation by TGF-β.

*Nossal:* The two observations may be related—it is possible that a major effect of the matrix proteins is to capture local growth factors.

## References

Chambard J-C, Pouysségur J 1988 TGF-β inhibits growth factor-induced DNA synthesis in hamster fibroblasts without affecting the early mitogenic events. J Cell Physiol 135:101–107

Klein G 1989 Reversion of tumorigenicity in an EBV-converted Burkitt's lymphoma line. In: Genetic analysis of tumour suppression. Wiley, Chichester (Ciba Found Symp 144) p 36–53

Silberstein GB, Daniel CW 1987 Reversible inhibition of mammary gland growth by transforming growth factor-beta. Science (Wash DC) 237:291–293

# Proviral position effects: possible probes for genes that suppress transcription

*J.A. Wyke, †J. Akroyd, *D.A.F. Gillespie, †A.R. Green and *C. Poole

**Beatson Institute for Cancer Research, Garscube Estate, Switchback Road, Bearsden, Glasgow, G61 1BD and †Imperial Cancer Research Fund Laboratories, St Bartholomew's Hospital, London, EC1A 7BE*

*Abstract.* Rous sarcoma virus, an oncogenic avian retrovirus, readily causes morphological transformation of chick cells, but in infected rat cells transformation is rare because proviral transcription is inefficient. This constraint is not due to a lack of positive transcriptional factors, or an excess of negative ones, but reflects the site of proviral integration in rat cell DNA. In most sites the provirus is almost invariably silent, in others it is correspondingly active, whilst in a third category expression fluctuates in concert with transitions in chromatin structure. Transcriptional fluctuations are mediated both by flanking cell DNA in *cis* and by *trans*-acting cell genes, suggesting that proviral position effects are sensors for genes that down-regulate transcription, perhaps by determining chromatin configuration. We have tried to identify such genes by gene transfer, karyology and insertional mutagenesis. The variable success of these three approaches indicates that the transcriptional down-regulator(s) need act only transiently. This is consistent with a function that operates in ontogeny or differentiation to down-regulate genes whose silence is then perpetuated by other means. The loss of such functions may predispose to neoplasia.

*1989 Genetic analysis of tumour suppression. Wiley, Chichester (Ciba Foundation Symposium 142) p 117–130*

Tumour suppressor genes have been identified largely by operational criteria and our knowledge of their functions is mostly rudimentary. It seems likely that they serve in the normal organism to promote development and differentiation, occupying niches in pathways of cellular information processing that are comparable to those in which proto-oncogenes are ensconced. One such niche is the regulation of transcription, and mechanisms that determine the accessibility of DNA to transcription factors, or affect the factors' activity, could well involve currently unknown genes whose ultimate phenotypic effects include tumour suppression.

We were drawn to this concept by studies on an apparently unrelated phenomenon, restriction of the activity of retroviral proviruses integrated in host cell DNA. In this article we summarize evidence that transcriptional control of proviruses can be a consequence of their site of integration, this modulation revealing host cell regulatory mechanisms that are hence amenable to analysis.

## The observations

Fig. 1 depicts the consequences of infecting either chick or rat cells with the transforming retrovirus, Rous sarcoma virus (RSV). In chick, the proviral long terminal repeat (LTR) efficiently drives expression in every cell, leading to transformation (mediated by the v-*src* oncogene) and virus replication. In Rat-1 cells, virus replication does not occur but appropriate RSV strains efficiently establish a proviral state. Although many of these proviruses are potentially active, the vast majority fail to cause cell transformation (rare 'delayed' transformants may appear in time). Moreover, the minority of cells that are transformed can segregate morphologically normal revertants which frequently retain a competent provirus, indicating that proviral expression can be modulated. Such modulation is displayed more dramatically by the high frequency suppression of the transformed phenotype in certain hybrids between RSV-transformed Rat-1 cells and normal cells of several types (Dyson et al 1982).

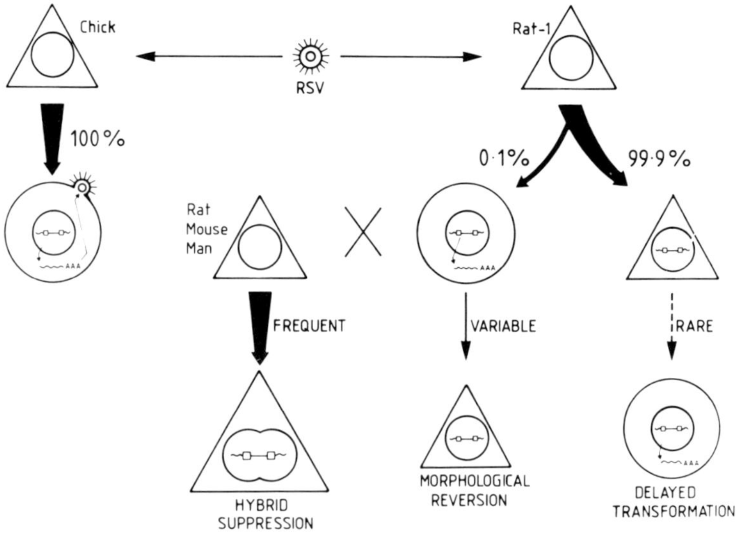


FIG. 1. The variable consequences of RSV infection of chick and Rat-1 cells. Cells depicted as triangular are phenotypically normal, round cells are transformed. Provirus transcription is invariably seen in transformed cells but virus replication occurs in chick cells only.

Proviral 'silence' is mediated at the level of initiation or elongation of transcription of the whole provirus (Akroyd et al 1987). Its occasional abrogation is not due to mutant viruses that are adapted to function in Rat-1 cells, to mutant cells that are particularly prone to transcribe the provirus, or to a physiologically variant, potentially saturable, subpopulation of cells (Wyke & Quade 1980). Indeed, in short term assays the RSV LTR functions efficiently in uninfected Rat-1 cells, in RSV-transformed derivatives and in cells containing a silent provirus, showing that none of these cells lacks factors required for proviral transcription (Akroyd et al 1987).

Further details came from studies on Rat-1 cells containing single identifiable proviruses, particularly those in which proviral expression fluctuates. Revertants of cells originally transformed by wild-type RSV can be superinfected by RSV carrying a temperature-sensitive v-*src* (Fig. 2). Transformants so obtained have a uniformly temperature-sensitive phenotype, showing that the superinfecting virus is expressed, whilst the original one remains silent. The only difference between these two infections is the time and site of proviral integration.

To rule out time-dependent variations, we exploited hybrid suppression of transformation. Proviral transcription in the transformed Rat-1 cell clone A23 is suppressed in a hybrid with uninfected Rat-1 cells, whereas the provirus in another Rat-1 cell clone, B31, continues to be expressed (Fig. 3; Dyson et al 1982). A mutation in the v-*src* gene of the B31 provirus has generated a phenotypically normal cell variant which still transcribes the provirus (Varmus et al 1981, H.E. Varmus, personal communication). This variant, when fused to clone A23, yields normal hybrids. Thus, the B31 cell, despite harbouring a transcribed provirus, can suppress the activity of the provirus in A23. We conclude that RSV proviral transcription is primarily dependent on its site of insertion in the Rat-1 genome. Examination of individual integrated proviruses has provided both circumstantial and direct support for this.

Firstly, duplications of proviral (and sometimes host) DNA were found adjacent to nine of 15 transcriptionally active proviruses in Rat-1 cells, and in all six cases for which sufficient data are available this has occurred immediately 5′ to an ostensibly intact provirus (Gillespie et al 1985). The absence of such duplications in the vicinity of 17 transcriptionally inert proviruses (Green et al 1986) suggests that they are not a mere concomitant of proviral integration but play a role in favouring transcription. We found (Levantis et al 1986) that the rearrangements do not result in a novel transcriptional unit with read-through of the adjacent provirus, a situation with precedents in some integrated murine leukaemia viruses (Barklis et al 1986). There are, however, other possible mechanisms (Levantis et al 1986) whereby the inhibition of proviral transcription by a nearby host negative regulatory element is abrogated.

(1) Where examined, the duplicated regions of provirus are those known

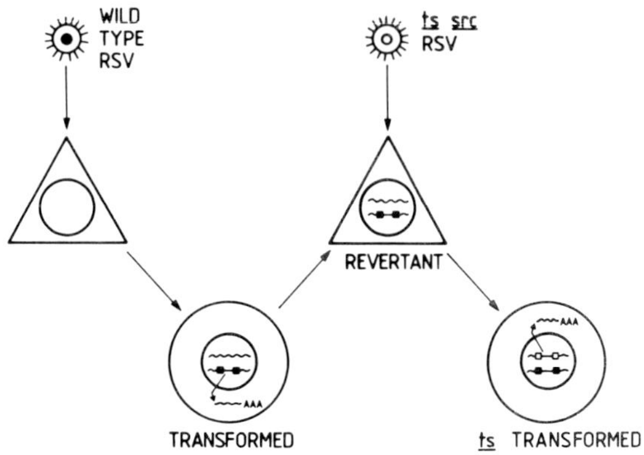


FIG.2. Proviral silence is position or time dependent.

to contain enhancers and it is conceivable that they augment the activity of the proviral LTR.

(2) Retroviral genomes contain more CpG doublets than eukaryotic DNA, so duplication results in a 'CpG-rich' island upstream of the provirus, a motif characteristic of the 5′ ends of many genes (Bird 1986).

(3) In at least one instance the duplication is transcribed at the same time as, and in the opposite orientation to, the provirus, another phenomenon found at the 5′ end of some genes (Efrat & Hanahan 1987).

(4) Finally, the duplication may simply distance the provirus from an upstream cell inhibitory element.

An inhibitory role for DNA flanking a provirus has been directly demonstrated. Cellular DNA from BB16, another Rat-1 clone containing an active provirus, efficiently transformed recipient NIH-3T3 mouse cells, whereas DNA from a daughter morphological revertant was inefficient. The transforming activity of the revertant DNA was reinstated by cleavage with a restriction enzyme that does not cut the provirus (Akroyd et al 1987). Flanking DNA for 3 kb on either side of the provirus was not cut by the enzyme, so it seems that an inhibitory element in cell DNA at least 3 kb from the integration site maintains BB16 proviral silence in NIH-3T3 cells. This conclusion is not as simple as it appears. Since we have no evidence that reversion in BB16, or in any other transformed rat clone, requires sequence alterations in the DNA flanking the provirus, it seems that sequences in *cis* to the revertant provirus maintain transcriptional silence, whilst the same sequences in its transformed progenitor are unable to suppress transcription. We tentatively conclude that the initiation and maintenance of RSV transcriptional silence in clone BB16 are phenomena separable by one (or both) of two reasons. 1) NIH-3T3 cells may have the capacity to perpetuate, but not initiate, transcriptional silence at the donor cell provirus locus. 2) Irrespective

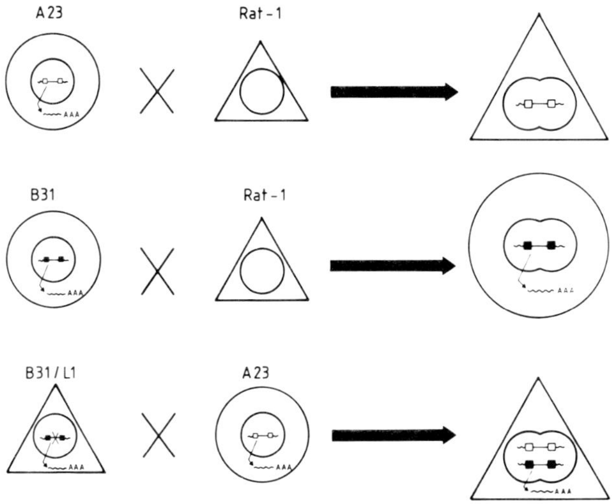


FIG. 3. Hybrid suppression of transformation appears site specific. A23 and B31 are two RSV-transformed Rat-1 clones. B31/L1 is a subclone of B31 rendered morphologically normal by a mutation in the proviral v-*src* gene.

of the abilities of the recipient cell, the *cis* element initiating transcriptional silence is further from the provirus than the element that perpetuates it, and is thus more often separated from the provirus at transfection.

Possible mechanisms are suggested by the observations that cessation of provirus transcription in both revertants and suppressed hybrids is accompanied by structural changes at the provirus loci (Dyson et al 1985, and references therein) which become 1) insensitive to nucleases, with a specific loss of hypersensitive sites in the 5′ LTR and *gag* gene, 2) markedly distanced from the nuclear scaffold, and 3) hypermethylated at a subset of *Msp*I/*Hpa*II (CCGG) sites. It is possible that the transitions in chromatin configuration play a role in initiating transcriptional suppression whilst cytosine methylation maintains it. Consonant with this is the observation that the methylation antagonist, 5-azacytidine, can induce retransformation of revertants in a dose-dependent manner, but with an efficiency that varies greatly between different proviruses. This suggests that cytosine methylation is a cause, rather than a consequence, of transcriptional silence, but one whose effect can only be relieved in the absence of overriding influences.

Position effects on transcription have also been invoked to explain the variable expression of proviruses introduced into embryonic and other stem cells (Jaenisch et al 1981, Magli et al 1987). Most information has come from studies on embryonal carcinoma (EC) cells, whose restriction of Moloney murine leukaemia (M-MuLV) provirus expression is comparable in extent to that exerted by Rat-1 over RSV, but interestingly for different reasons (Fig.

4). In EC cells the M-MuLV LTR is inactive, probably because a *trans*-acting negative regulatory element acts on its enhancer (Gorman et al 1985). This inhibition is relieved by proviral mutation or proximity to cellular positive regulatory elements present at a very small number of integration sites. In some cases transcription originates in cell sequences and reads through the provirus, whilst in one instance proviral transcription initiates correctly in the viral LTR, perhaps reflecting the operation of cell promoter and enhancer elements, respectively (Barklis et al 1986).

In contrast, the unintegrated RSV LTR functions efficiently in Rat-1 cells but on integration is susceptible to *cis*-acting cellular 'silencers'. The prevalence of this phenomenon suggests that these silencers are widely distributed throughout the genome or else can act over very long distances. Rare proviral expression may occur by one of several routes. 1) Integration may occur into a favoured site beyond the influence of silencers. Proviruses in such sites would not be subject to transcriptional modulation and some examples fit this category. 2) The provirus may become physically or functionally distanced from the silencer as a result of upstream rearrangements (see above). 3) Mutations may make the provirus refractory to silencer effects. 4) The silencer's action may be modulated, by mutation or by down-regulation of a *trans*-acting inhibitor; the phenomenon of hybrid suppression predicts the existence of such an inhibitor.

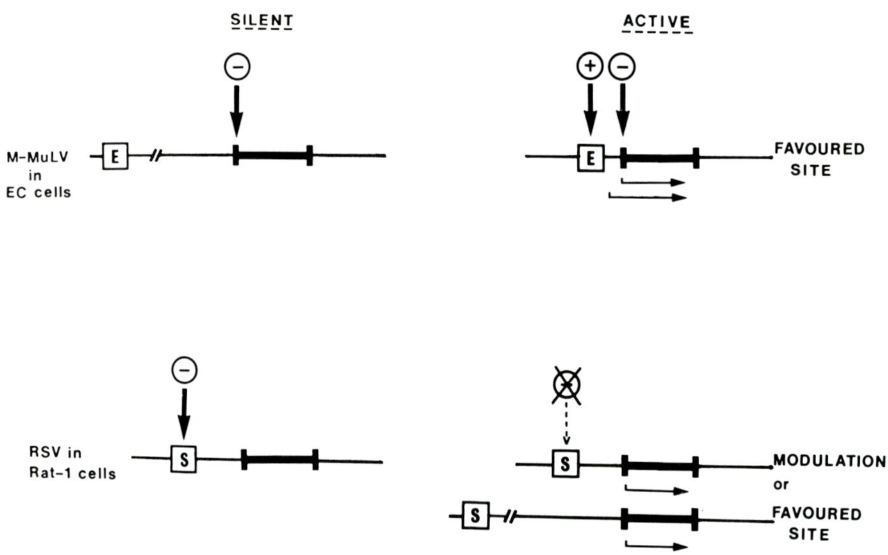


FIG. 4. Molecular mechanisms for chromosome position effects controlling M-MuLV and RSV proviral transcription in EC and Rat-1 cells, respectively. S, putative cell silencer; E, putative cell enhancer; — and +, *trans*-acting inhibitory and stimulatory molecules, respectively; thick lines, proviruses; thin lines, cell DNA.

In summary, the model suggests that proviral expression is the net result of the activity of two regulatory units — the proviral LTR and *cis*-acting cellular elements — both of which are potentially influenced by positive and negative *trans*-acting factors. Proviral rearrangements may interpose a third regulatory unit, whose modulation could explain how transcriptional silence can be imposed on rearranged proviruses without altering the rearrangement. Regulation through the LTR predominates in M-MuLV infection of EC cells, whereas flanking cell sequences are the prime determinants of RSV expression in Rat-1. It follows that RSV proviruses detect otherwise inaccessible cell controls; *cis*-acting silencers and *trans*-acting suppressors of transcription.

**Investigating the mechanisms**

A complete description of positional effects on proviral transcription must unravel both *cis* and *trans* influences, and cells in which transcriptional fluctuations can be selected or induced provide the most promising tools. The search for *cis*-acting silencers in the DNA that flanks proviruses is potentially laborious, requiring a judicious choice of provirus and assay system for silencer effects. Evidence for *trans*-acting influences comes mainly from hybrid suppression of transcription, and this phenomenon provides a basis for direct attempts to isolate a suppressor gene.

Fusion of phenotypically normal mouse or human cells to a RSV-transformed Rat-1 clone containing a single provirus (B2*LA*8) suppresses provirus transcription. It is thus possible that suppression can be induced by DNA transfer from the same normal cells. We therefore co-transfected mouse genomic DNA and pSV2neo (encoding G418 resistance) into B2*LA*8, finding morphologically normal recipients among the drug-resistant colonies. Their DNA, in a second round of transfection, again yielded morphologically normal B2*LA*8 that had acquired mouse sequences, suggesting that a mouse suppressor gene has been transmitted through both rounds of transfection (Green & Wyke 1988). Unfortunately, there is an unacceptably high incidence of normal colonies in control transfections. We were further concerned that transcriptional suppression might require multiple genes or a large gene that would not be revealed by simple DNA transfer.

Accordingly, we analysed hybrids between B2*LA*8 and human cells (both foreskin fibroblasts, HFF, and peripheral blood mononuclear cells, PBMC) in the correct anticipation that they would be phenotypically unstable and would contain variable complements of human chromosomes. This phenomenon offered several approaches to identifying transcriptional suppressors, as outlined in Table 1.

Since hybrids were phenotypically unstable only early after isolation, karyological analysis compared sibling transformed and normal hybrids rather than different generations in a lineage. Summaries of results (Tables 2

**TABLE 1 Approaches to identifying transcriptional suppressors in hybrid cells between RSV-transformed Rat-1 and human cells**

| *Approach* | *Requirement* | *Problems* |
|---|---|---|
| Karyology | Unstable hybrid phenotype | Possible transient effect of suppressor gene |
| Insertional mutagenesis | Stable hybrid phenotype | |
| Gene transfer<br>DNA<br>Chromosomes<br>Microcells | Stable recipient (B2*LA*8) phenotype | Concern about polygene effects decreases with increasing size of transgene, but so does definition |

and 3) show that no single chromosome is present in all normal hybrids and absent in all transformed siblings. Interpretation of karyological data is difficult but these results suggest one of two conclusions; 1) transcriptional suppression requires the continued concerted action of genes on more than one chromosome, or 2) the suppressor gene or genes need act only transiently after hybrid formation.

The subsequent phenotypic stability of normal hybrids suggested a direct way to identify a suppressor gene — insertional mutagenesis by a non-transforming retrovirus leading to gene inactivation and consequent cell transformation. A hybrid between B2*LA*8 and human PBMC (8/4/4A) contains single copies of a small number of human chromosomes and at least 20% of the hybrid cells retain the suppressor gene, as judged by their ability to suppress transcription when fused with the parent cell line, B2*LA*8 (Table 4). However, insertion of approximately $4 \times 10^6$ genomes of the retrovirus ZIP(neo)SV(X)1 into 8/4/4A failed to transform it. This is difficult to reconcile with suppression requiring the concerted action of several genes but it is consistent with the idea that the suppressor gene need act only transiently.

**TABLE 2 Karyology of B2*LA*8 × human foreskin fibroblast hybrids[a]**

Human chromosomes *common* to all six *normal* hybrids:

6, 7, 11, 12, 22, X (mean 13 chromosomes per hybrid)

Human chromosomes in four *transformed* segregants:

1) 3, 4, *6*, *7*, 8, *11*, 13, 15, 16, 17, 19, 20, 21
2) 3, 4, 5, *6*, *7*, 8, *11*, *12*, 15, 16, 17, 20, 21, *22*, *X*
3) 4, 5, *6*, *7*, *11*, *12*, 15, 16, 17, 20, 21, *22*, *X*
4) 2, 3, 4, *6*, 10, *12*, 13, 17, 19, 20, 21, *X* (no data on 7, 8, 22)

All transformed segregants have lost chromosomes, but not necessarily those from the common group. Those chromosomes common to all normal hybrids are italicized in the transformed segregants.

[a] We are grateful to the laboratories of Drs S Povey, D Sheer and B Young for karyological analyses.

**TABLE 3 Karyology of B2*LA*8 × human peripheral blood mononuclear cell hybrids**

Human chromosomes *common* to four normal hybrids:
11, 17, X (mean 14 chromosomes per hybrid)

Human chromosomes in four *transformed* segregants:

1) *X*
2) 5, 21
3) 3, 6, 7, 8, 10, 13, 16, *17*, 18, 20, 22, *X*
4) 6, *X*

Further support for this concept comes from the distinction between initiation and maintenance of transcriptional silence in revertants (see above); a transient requirement for the initiating event is implicit in such two component control. This reasoning strongly suggests that gene transfer provides the best means to identify suppressor genes and to this end a pool of 8/4/4A cells tagged with ZIP(neo)SV(X)1 should prove a useful resource (Table 4).

## The significance of position effects

Why is position-dependent suppression of RSV transcription so much more prevalent in Rat-1 cells than in chick cells (Fig. 1)? One possibility, mooted by zur Hausen (1986), is that eukaryotic cells possess an intracellular surveillance mechanism that down-regulates expression of foreign inserted DNA. If so, then the RSV provirus is recognized as foreign by Rat-1 cells but not by chick, and it follows that some portion of the virus genome must elicit this response at some, but not all, integration sites. Rosl et al (1988) suggest that a factor specific to non-tumorigenic cell hybrids may down-regulate transcription of human papillomavirus type 16 in HeLa cells and may account, at least in part, for the tumour suppressing capacity of human chromosome 11. It is intriguing that human chromosome 11 is retained by all normal hybrids between B2*LA*8 and human cells (Tables 2 and 3), but when backcross hybrids between 8/4/4A and B2*LA*8 are separated on the basis of chromosome 11 specific markers, there is no correlation between the presence of this

**TABLE 4 Characteristics of B2*LA*8 × peripheral blool mononuclear cell hybrid, 8/4/4A**

Normal morphology, very stable

Diploid number of rat chromosomes and single copies of human chromosomes 4, 11, 14, 17, 20, 21, X (plus fragments)

8/4/4Aneo provides pool of 17 000 colonies tagged with ZIP(neo)SV(X)1

At least 20% of cells retain suppressor gene

Forms micronuclei in response to colcemid

chromosome and a normal phenotype (A.R. Green, unpublished data).

A second possibility is that RSV proviruses reflect pre-existing regulatory phenomena at their site of integration. The extreme contrast in their behaviour in Rat-1 and chick cells may be because the interplay of regulatory elements in Rat-1 (see above) is tilted towards transcriptional suppression. Alternatively, the sites of integration in the two species may differ, since it seems that insertion sites are multiple but not random (Shih et al 1988). If the suggested preference for integration in regions of 'open' chromatin configuration (Vijaya et al 1986, Rohdewohld et al 1987) is not followed by RSV in Rat-1 cells, then most silent proviruses may simply be inserted in chromatin inaccessible to transcription factors.

In conclusion, the majority of silent RSV proviruses in Rat-1 cells may reflect nuances of proviral insertion or LTR potency, or they may result from hitherto unsuspected intracellular defence mechanisms. Comparisons of integration sites in rat and chick cells, before and after proviral insertion, should distinguish these possibilities. The minority of proviruses that show fluctuations in expression appear to detect separable mechanisms for initiation and maintenance of transcriptional silence. These processes could determine major shifts in gene expression during development and differentiation and their perturbation might contribute to neoplastic growth and behaviour.

## References

Akroyd J, Fincham VJ, Green AR, Levantis P, Searle S, Wyke JA 1987 Transcription of Rous sarcoma proviruses in rat cells is determined by chromosomal position effects that fluctuate and can operate over long distances. Oncogene 1:347–354

Barklis E, Mulligan RC, Jaenisch R 1986 Chromosomal position or virus mutation permits retrovirus expression in embryonal carcinoma cells. Cell 47:391–399

Bird AP 1986 CpG-rich islands and the function of DNA methylation. Nature (Lond) 321:209–213

Dyson PJ, Quade K, Wyke JA 1982 Expression of the ASV *src* gene in hybrids between normal and virally transformed cells: specific suppression occurs in some hybrids but not others. Cell 30:491–498

Dyson PJ, Cook PR, Searle S, Wyke JA 1985 The chromatin structure of Rous sarcoma proviruses is changed by factors that act in *trans* in cell hybrids. EMBO (Eur Mol Biol Organ) J 4:413–420

Efrat S, Hanahan D 1987 Bidirectional activity of the rat insulin II 5′-flanking region in transgenic mice. Mol Cell Biol 7:192–198

Gillespie DAF, Hart K, Wyke JA 1985 Rearrangements of viral and cellular DNA are often associated with expression of Rous sarcoma virus in rat cells. Cell 41:279–287

Gorman CM, Rigby PWJ, Lane DP 1985 Negative regulation of viral enhancer in undifferentiated embryonic stem cells. Cell 42:519–526

Green AR, Wyke JA 1988 Integrated proviruses as probes for chromosomal position effects in mammalian cells and their hybrids. Cancer Surv 7:335–351

Green AR, Searle S, Gillespie DAF, Bissell M, Wyke JA 1986 Expression of integrated Rous sarcoma proviruses: DNA rearrangements 5′ to the provirus are common in transformed rat cells but not seen in infected but untransformed cells. EMBO (Eur Mol Biol Organ) J 5:707–711

Jaenisch R, Jahner D, Nobis P et al 1981 Chromosomal position and activation of retroviral genomes inserted into the germ line of mice. Cell 24:519–529

Levantis P, Gillespie DAF, Hart K, Bissell MJ, Wyke JA 1986 The control of expression of an integrated Rous sarcoma provirus in rat cells: studies on the role of 5′ genomic duplications reveal unexpected patterns of gene transcription and its regulation. J Virol 57:907–916

Magli M-C, Dick JE, Huszar D, Bernstein A, Phillips RA 1987 Modulation of gene expression in multiple hematopoietic cell lineages following retroviral vector gene transfer. Proc Natl Acad Sci USA 84:789–793

Rohdewohld H, Weiher H, Reik W, Jaenisch R, Breindl M 1987 Retrovirus integration and chromatin structure: Moloney murine leukemia proviral integration sites map near DNase1-hypersensitive sites. J Virol 61:336–343

Rosl F, Durst M, zur Hausen H 1988 Selective suppression of human papillomavirus transcription in non-tumorigenic cells by 5-azacytidine. EMBO (Eur Mol Biol Organ) J 7:1321–1328

Shih C-C, Stoye JP, Coffin JM 1988 Highly preferred targets for retrovirus integration. Cell 53:531–537

Varmus HE, Quintrell N, Wyke JA 1981 Revertants of an ASV-transformed rat cell line have lost the complete provirus or sustained mutations in *src*. Virology 108:28–46

Vijaya S, Steffen DL, Robinson HL 1986 Acceptor sites for retroviral integrations map near DNase1-hypersensitive sites in chromatin. J Virol 60:683–692

Wyke J, Quade K 1980 Infection of rat cells by avian sarcoma virus: factors affecting transformation and subsequent reversion. Virology 106:217–233

zur Hausen H 1986 Intracellular surveillance of persisting viral infections. Lancet ii:489–491

## DISCUSSION

*Stanbridge:* John, it might be a little dangerous to rely solely on antigen expression as an indicator of the presence or absence of a given chromosome in your backcross hybrids between 8/4/4A and B2*LA*8. Have you used DNA probes that map to chromosome 11p and 11q to see if there is any genetic information from chromosome 11 in the supposedly negative cells?

*Wyke:* No, we haven't. We do appreciate this problem, but the distribution was so unbiased in the two populations that we felt chromosome fragmentation was an unlikely explanation, unless of course the fragmentation had occurred in the original hybrid that we used to make the fusion. There are fragmented chromosomes in 8/4/4A; for example, a fragment of chromosome 14. I think the best way to answer your point would be to repeat fusions at other times with other hybrids.

*Stanbridge:* Yes, but before you conclude that chromosome 11 is not implicated in transcriptional suppression, it would be important to use DNA probes.

*Wyke:* Thank you, yes. So far the theory is not sustainable, but that does not mean it may not be ultimately correct.

*zur Hausen:* John, have you tried to activate your down-regulated provirus by treating your cells with, for example, 5-azacytidine?

*Wyke:* I haven't done it with hybrids, but if one treats revertant cells with 5-azacytidine one can activate the provirus in a dose-dependent manner. This effect shows extreme variability between revertant clones. This could be consistent with an idea that, if methylation is involved in the maintenance of proviral inactivity, then the effect of 5-azacytidine might depend on whether or not the imposing mechanism is still functioning.

*Varmus:* Your results on down-regulation suggest that the propensity to down-regulate is sequence determined or site determined. But your lab and mine have cells in which provirus transcription is always maintained and others in which it can be turned off. Yet your transfection experiments suggest that modification of DNA is likely to determine whether or not that DNA is expressed when introduced into cells. If you separate a revertant provirus from its flanking DNA, you may lose the determinant for whether or not the provirus is expressed after transfection. Have you tried to map the sequences that determine whether or not the modification will occur, by cloning DNA from different sites and doing mix-and-match experiments?

*Wyke:* We have tried to do this. I alluded to this approach of looking for *cis*-regulatory elements and I mentioned that it was laborious and required the right choice of recipient cell to detect 'silencer' effects. Our problem so far has been to identify a recipient in which we know that we can turn off transcription of the 'active form' of the provirus. In the experiment with the clone BB16, NIH-3T3 cells are not a suitable recipient for these sorts of mix-and-match experiments. We are now making a number of hybrids between our transformed BB16 line and various normal lines to identify a hybrid in which we can turn off the expression of the virus. We have already made cosmid clones of the proviral locus in BB16, so once we have a suitable recipient we can put the flanking sequences into that.

*Stanbridge:* There is an interesting apparent dichotomy between your studies and those of Dr Noda, who took cells transformed by murine sarcoma virus and obtained flat revertants which continued to express the viral *src* oncogene. Do your revertants exclusively shut off *src* expression, John, or do you see different categories of revertants—those that down regulate and those that don't?

*Wyke:* I have never found a revertant in which *src* expression is retained. Harold (Varmus) has described one and I have tried very hard to do the same thing, working on the hypothesis that in a cell transformed by multiple proviruses you are not going to get individual shut-off at a determinable frequency. In the single revertant I obtained from such a cell line, the data suggest that the cell is a mutant but a mutant that has acquired the ability to down regulate the activity of the viral LTR. The problem in our system is that the modulation of proviral transcription occurs at such a level that the sorts of revertants Dr Noda has found would only be a small minority of our revertants. The B31 line, in which transcriptional down-regulation has never been observed, perhaps because the provirus is in a particularly favoured site in the cell, would be the candidate of choice to search for revertants resembling Dr Noda's. It was, in fact, the starting point for Harold's experiments. In the past Harold has also found a number of viral mutants as a result of using this line.

*Adams:* How many of your results could be explained simply by the viral DNA containing a sequence that attracts methylation, which then shuts down viral expression? If that were the case, only in certain insertion sites would the provirus escape inactivation.

*Wyke:* In a sense that's another way of saying the same thing: the question is, what attracts methylation and why? One possibility is that, although I presuppose insertion is random, we don't know that is always the case. We know that insertion is multiple, but it is possible that in the majority of cells the viruses are silent because they do insert at random, whereas viruses that are adapted to their host cell insert preferentially into active sites. It could be that a lot of these viruses that are silent *ab initio*, have, for some reason, integrated into inactive chromatin. This is the reason for looking further at insertions where modulation occurs. In those cases you know that you can get expression at the insertion site and its subsequent modulation implies an interesting mechanism.

*Noda:* Is there any possibility that in the experiments on cell hybridization you are looking at two things—suppression of the expression of v-*src* and suppression of the transformed phenotype? Did you look at the expression of v-*src* in those hybrids?

*Wyke:* We know that we are looking at suppression of proviral expression because whenever we look at a hybrid that is phenotypically normal we know that the virus is not being transcribed. We know by nuclear 'run on' experiments that regulation is occurring at the level of initiation or elongation of transcription. What I cannot say is whether there is an additional effect on the transformed phenotype. When you make hybrids between cells of different phenotype, you do get *trans*-regulation of the differentiated phenotypes, as exemplified by the work of Mary Weiss and her collaborators on fusions between hepatocytes and other differentiated cells (Weiss et al 1988). I cannot rule out the possibility that, if I could maintain viral expression, there might be some other effect operating at the post-transcriptional level. But I know transcription down-regulation operates and that is sufficient to explain what I see.

One reason that we did a lot of experiments with B2*LA*8, a cell line in which the virus contains a temperature-sensitive mutation, was simply to test that whenever one did get, for example, segregation or spontaneous transformation, we could easily check that it was due to re-expression of the viral gene and not to some unrelated spontaneous event.

*Varmus:* You could test Dr Noda's idea by superinfecting any hybrid with murine leukaemia virus carrying *src*.

*Wyke:* Yes, indeed. We have transformed revertants by superinfection with that virus carrying *src*. Most of that's been done with the revertant which contains multiple proviruses.

## Reference

Weiss MC, Cassio D, Sellern CH 1988 Cell differentiation: contributions of somatic cell genetics. Cancer Surv 7:294–302

# General discussion IV

## Methods for isolating genes that act in a recessive manner in tumorigenesis

*Varmus:* There has been a lot of discussion of the phenomenology of the, as yet unnamed, genes that act in a recessive fashion during oncogenesis. The task still facing us is to isolate these genes by molecular cloning. One strategy is that which Bob Weinberg used to isolate the *Rb* gene: luck! Another is that which Nick Hastie and others have been following to isolate the Wilms' locus, namely a long march along the chromosome. A third possibility is to use functional assays as tests for the gene one would like to clone, for example, a transfection assay in which one looks for the loss of some transforming property of a cell that lacks a functional suppressor gene. This is obviously difficult because it requires an assay for loss of function. We've tried to invert this approach by using a Wilms' tumour cell line which has lost the ability to grow in soft agar as a result of chromosome transfer, as reported by Eric Stanbridge (see Stanbridge, this volume). We used a feline retrovirus, RD114, that grows well in human cells, to inactivate by insertion the gene believed to be responsible for suppression of the ability of the cells to grow in soft agar. Thus an insertion mutation will allow the cell to grow in soft agar and give a selectable phenotype: the provirus is then used as a tag to isolate the gene. This strategy has been tried unsuccessfully in our lab by Titia de Lange. We have collaborated with Eric to use some of the lines that he has developed, but the phenotype for which we would select has not been tight enough to allow us to carry out the assay.

An additional strategy involves differential cDNA cloning. Titia de Lange has developed a clever general strategy, although it is not clear whether her experiments with Wilms' tumour have produced useful clones yet. The basis for the approach is evident in the work described by Bob Weinberg. If one looks at a series of retinoblastomas, one finds that a certain proportion of tumours (up to 50%) lack an *Rb* gene transcript. It is possible that transcripts in a series of Wilms' tumours might be identical except for the presence or absence of the transcript from the suppressor gene. The idea is therefore to isolate cDNAs that are specific for that one transcript. The dangerous assumption is that the tumours are going to have the same transcriptional phenotype, except for the expression of the gene(s) of interest.

We made a cDNA library from an arbitrarily selected series of six Wilms' tumours, called A–F, and the library was screened with a series of subtractive probes. To make these probes, labelled cDNA was prepared from RNA from tumour A and subtracted individually with unlabelled RNA from the other tumours, B–F. To deal with the possibility that A was a tumour which didn't express the Wilms' locus, labelled cDNA was made from B,C,D,E and F RNA and subtracted with RNA from A. There should then be labelled cDNA in one of these pools that is specific for the transcript of interest, assuming that one of the six tumours differs from the others with respect to the presence or absence of the Wilms' tumour transcript.

The six probes were then used to screen many thousands of colonies from the original cDNA library. It was possible to isolate from roughly 50000 plaques 30 that appeared to be specific for two of the probes. In other words, it appeared that two of the tumours differed from the other four with respect to 30 clones, showing that the transcriptional programme of the tumours was different. Are these 30 clones from the Wilms' locus? In 29 of the 30 cases the answer is no. Titia did a simple screen to ask whether the clones are derived from genes that map to chromosome 11: hybridization to filters with restriction digests of human DNA, mouse DNA, and DNA from a mouse line that carries human chromosome 11. 29 clones failed to give the bands that one would expect if chromosome 11 were the source of the clones.

The exception is a clone called 42, which does hybridize to human chromosome 11. The next step is to ask whether this could be the Wilms' gene. At the moment the evidence is against it. Analysis of a large number of Wilms' tumours suggests that the tumours don't contain a rearranged gene that is detected by the cDNA clone. *In situ* hybridization by Michelle Le Beau suggests that the gene maps on the telomeric side of 11p13, closer to p14 or p15. We are pursuing the possibility that there might be two Wilms' loci. Web (Cavenee) and his colleagues are looking at some of their material for the possibility that this clone maps to the region involved in embryonic rhabdomyosarcoma and other syndromes, e.g. Beckwith-Weidemann, that may be associated with Wilms' tumour.

*Bodmer:* This is obviously a strategy that one can follow in many different ways. With respect to colorectal carcinoma and polyposis, we have been following an analogous strategy. I don't have any results, but there is evidence to suggest differential expression of a particular gene in colonic epithelial cells. So you can look for messages that are relatively prevalent in colonic epithelial cells; we use a very simple screen. The number of messages that are likely to be differentially expressed in epithelial cells that map not only to chromosome 5 but to the deletion position would be very small. The limitation we believe is the number of clones that you can screen with respect to the mapping, and we place the greatest emphasis on the mapping of cDNAs. If we could saturate the deletion region with all possible cDNAs for epithelial cells, we would be fairly sure that the relevant gene for polyposis would be in that region.

*Varmus:* The difficulty there is that you may begin with too many RNAs.

*Bodmer:* No, that's where the genetic technique is so powerful, because the number of relevant messages in that region is very small. The real issue is how many cDNAs can you screen and how stringently do you define differential expression. For instance, we have reason to believe that there is no expression of the familial adenomatous polyposis gene in lymphoid cells because individuals with the disease do not get lymphoid tumours. We are more hesitant with fibroblasts, since desmoids and osteomas occur, so you can get mesenchymal tumours.

*Adams:* Wouldn't that reasoning have led you astray with the *Rb* gene, because that is widely expressed?

*Bodmer:* Yes, but not if your main screen is done by genetics.

*Weinberg:* I think you would be on thin ice, if you assumed first that the gene which predisposes to colorectal carcinoma is expressed in a narrow range of tissue types, and secondly that normal colonic epithelium has a range of messenger RNA expression which closely approximates to that of colon carcinoma. If I understand your counter screen correctly, you are depending on that assumption?

*Bodmer:* No, we are depending on looking at the chromosome 5q21 region. I believe that, notwithstanding the retinoblastoma story, there is a good chance that the gene we find will be differentially expressed in epithelial cells.

*Varmus:* We assumed that Wilms' tumours have fairly similar transcriptional programmes. Making predictions like that is very dangerous: we have shown that there are at least 29 genes that differentiate two classes of Wilms' tumours.

## Reference

Stanbridge EJ 1989 A genetic basis for tumour suppression. In: Genetic analysis of tumour suppression. Wiley, Chichester (Ciba Found Symp 142) p 149–165

# Flat revertants of EJ human bladder carcinoma cells show two different mechanisms of reversion

H. Paterson, G. Mbamalu and C.J.Marshall

*Institute of Cancer Research, Chester Beatty Laboratories, Fulham Road, London SW3 6JB, UK*

*Abstract.* To investigate the way in which *ras* proteins cause transformation, we have isolated revertants from human tumour cell lines which contain transforming *ras* genes. Two types of revertant have been isolated from the human fibrosarcoma cell line, HT1080. One class has normal and mutant alleles in a ratio of 2:1, compared to 1:1 in the parental cells, showing that reversion can be a dosage phenomenon. The other class has lost the transforming allele. All the HT1080 revertants isolated can be re-transformed by transforming *ras* proteins. To test whether reversion is due to a change in the relative amounts of normal and mutant proteins, or to a reduction in the absolute amount of the transforming protein, mixtures of the purified proteins were microinjected into 208F (Rat-1) cells, chosen because they are less sensitive to transformation by p21*ras*. Normal H-*ras* p21 was unable to suppress the transforming effects of the mutant *ras* protein when co-injected at up to ninefold excess. Revertants of EJ human bladder carcinoma cells were of two types: one was sensitive to re-transformation by oncogenically activated *ras* proteins, the other was not. The EJ revertants that are resistant to re-transformation fall into two classes, since hybrids of one revertant with the parental EJ cells are non-transformed, whereas hybrids of another revertant with the parental cells are transformed.

*1989 Genetic analysis of tumour suppression. Wiley, Chichester (Ciba Foundation Symposium 142) p 134–148*

The *ras* oncogenes appear to play a significant role in human carcinogenesis, because a number of different malignancies contain point-mutated *ras* genes (for review see Bos 1988). These point mutations lead to the oncogenic activation of the *ras* proteins by reducing the intrinsic GTPase activity of the protein. Although a great deal is known about the pattern of *ras* gene activation in different malignancies, the details of how *ras* proteins transform cells are much less well understood. It is clear from a variety of experiments that in order to transform, the *ras* proteins must be localized in the membrane (Willumsen et al 1984). However, the molecules with which *ras* interacts are less well defined. Trahey & McCormick (1987) have defined a GTPase-activating protein (GAP) which interacts with normal *ras* proteins to greatly elevate GTPase activity. Genetic experiments suggest that this protein may in fact be the protein with which both normal and mutant p21 *ras* interact to produce their effects. Mutations defined as being in the effector domain because they destroy the transforming properties of p21*ras*, also destroy the ability of normal p21*ras* to interact with GAP (Calés et al 1988, Adari et al 1988). The biochemical consequences of the *ras*-GAP interaction are unknown. Circumstantial evidence suggests that normal p21*ras* may be part of an intracellular signalling system involved in growth control (Mulcahy et al 1985). This signalling system may involve the turnover of inositol lipids (Wakelam et al 1986, Fleischman et al 1986, Hancock et al 1988). Activation of protein kinase C appears to be an important component of the transforming effects of p21*ras*, since introduction of p21*ras* into quiescent cells leads to a rapid activation of protein kinase C (Morris et al 1989) and p21*ras*-induced DNA synthesis requires the activation of protein kinase C (Lacal et al 1987, Morris et al 1989). Recent results from our laboratory show that p21*ras* may also activate protein kinase C-independent events, because morphological transformation stimulated by p21*ras* protein does not require functional protein kinase C (unpublished results of A. Lloyd, J. Morris, A. Hall and C. Marshall). There is clearly, therefore, still a great deal to be learned about the interactions of p21*ras* proteins with other components of the cell.

To approach this question we have isolated revertants from human tumour cell lines which contain transforming *ras* genes. By this means we hope to understand how p21*ras* transforms cells and to identify genes which encode products that interact with *ras* proteins. Revertants isolated from the human fibrosarcoma cell line, HT1080, which contains one normal N-*ras* allele and one N61 lysine substitution, show that reversion to the non-transformed phenotype can occur by changes in the dosage of the transforming allele (Paterson et al 1987). A further revertant we have recently isolated from HT1080 has completely lost the N-*ras* transforming allele. All of these HT1080 revertants can be re-transformed by transforming *ras* protein; however, the revertants to be described here, which were isolated from the human bladder carcinoma cell line EJ/T24, are resistant to re-transformation by p21*ras*.

## Materials and methods

### *Isolation of revertants*

Revertants were isolated from mutagenized parental populations of EJ/T24 cells using the low serum 'suicide' selection protocol previously described (Paterson et al 1987).

*Cell culture*

Cells were routinely cultured in Dulbecco MEM supplemented with 10% fetal calf serum. Cell fusions were carried out using PEG1500, as previously described (Marshall & Sager 1981). Hybrids were selected in media containing 1.8 mg/ml G418, 0.25 mg/ml hygromycin B. For assays of anchorage-independent growth, cells were seeded in 0.3% agar (Bacto) over a 0.5% agar base in medium supplemented with 3 ng/ml epidermal growth factor. Calcium phosphate-mediated transfection was carried out essentially as described by Wigler et al (1978).

*Microinjection of p21ras proteins*

p21*ras* proteins were purified from *E. coli* expression systems, as described by Trahey et al (1987). Purified proteins were microinjected into the cytoplasm of quiescent cells over a range of concentrations. The cells were then cultured and their behaviour monitored by time-lapse video recording.

**Results**

*Does normal p21ras suppress the activity of mutant ras?*

Our previous work with the revertants of HT1080 showed that reversion occurred because the tetraploid revertants contained only one copy of the N-*ras* transforming allele and two copies of the normal N-*ras* allele, compared to the 1:1 ratio in the parental cells (Paterson et al 1987). Reversion could therefore be due to a change either in the absolute amount of the transforming protein or in the ratio of normal and transforming protein. Situations have been described in which there appears to be selection against expression of the normal *ras* allele (e.g. Capon et al 1983), suggesting that the normal p21*ras* may exert an inhibitory effect on transformation by p21*ras*. However, repression of the activity of the transforming N-*ras* protein by the normal N-*ras* protein does not seem to be the mechanism of reversion in the HT1080 revertants, because we were unable to revert HT1080 cells either by microinjection of normal N-*ras* p21 protein or by transfection with plasmids expressing normal N-*ras* p21 (Paterson et al 1987). The possibility remains, however, that because HT1080 cells are malignant, the sensitivity to suppression by the normal p21*ras* proteins has been lost. To investigate whether the presence of normal p21*ras* proteins can suppress the transforming activity of mutant *ras* proteins, we therefore co-injected mixtures of purified proteins into 208F cells. These cells, like the parental Rat-1 cells from which they were derived, are resistant to transformation by overexpression of normal p21*ras* proteins (Ricketts & Levinson 1988). These cells therefore appear to be less sensitive

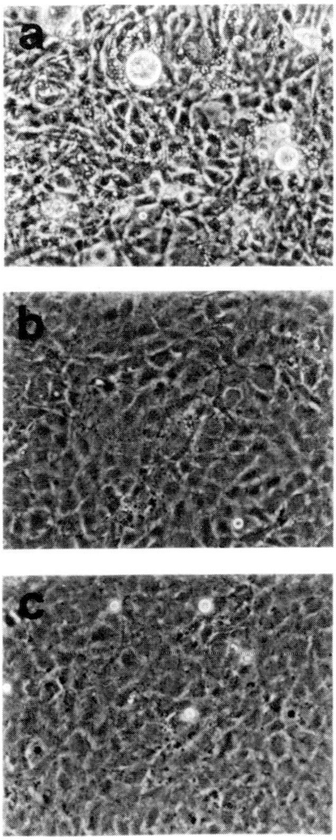

FIG. 1. Microinjection of normal or oncogenically activated p21*ras* proteins into 208F rat cells. (a) 1 mg/ml val[12]H-*ras* p21. (Transformed) (b) 10 mg/ml normal H-*ras* p21. (Flat) (c) Uninjected. (Flat)

to transformation by p21*ras* and provide a more appropriate substrate to test for suppressive effects of the normal proteins.

Quiescent confluent 208F cells were microinjected with p21*ras* proteins and then studied by time-lapse video microscopy. Fig. 1 shows the appearance of cells 48 hours after microinjection. 208F cells were transformed by microinjection of 1 mg/ml val[12]H-*ras* p21 (Fig. 1a) and appear to be less sensitive to p21*ras* than NIH-3T3 cells, which can be transformed by microinjection of 0.3 mg/ml val[12]H-*ras* p21 (data not shown). Doses of normal H-*ras* p21 of up to 10 mg/ml fail to transform 208F cells (Fig. 1b), whereas NIH-3T3 cells are transformed by injection of the same protein preparation at 3 mg/ml (data not shown).

Fig. 2 shows that when val[12]H-*ras* p21 at 1 mg/ml was mixed with varying concentrations of normal H-*ras* p21 up to 9 mg/ml and then microinjected into the 208F cells, transformation still occurred. Normal H-*ras* p21 therefore

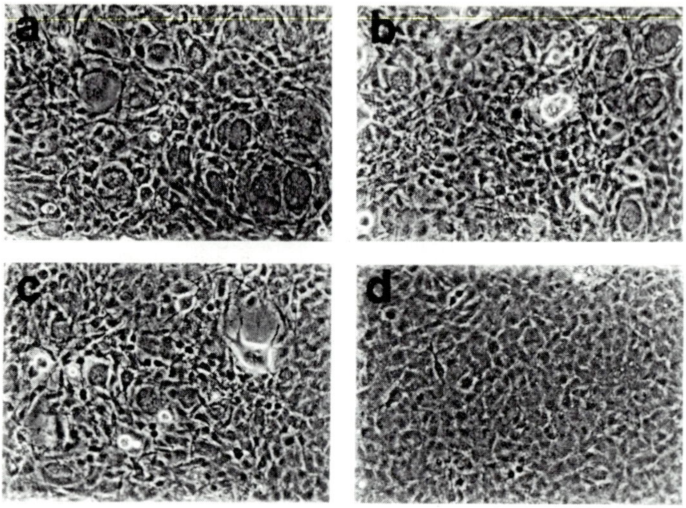

FIG. 2. Microinjection of mixtures of normal and oncogenically activated p21*ras* proteins into rat 208F cells. (a) 1 mg/ml val$^{12}$H-*ras* only. (Transformed) (b) 1 mg/ml val$^{12}$H-*ras* + 2 mg/ml normal H-*ras*. (Transformed) (c) 1 mg/ml val$^{12}$H-*ras* + 9 mg/ml normal H-*ras*. (Transformed) (d) Uninjected (Flat).

appears to be unable to suppress the transforming effects of val$^{12}$H-*ras* p21. Similar results have been found by Ricketts & Levinson (1988), who demonstrated that a Rat-1 cell line overexpressing normal H-*ras* p21 was as sensitive as the parental Rat-1 cells to transformation by transfection with a plasmid expressing val$^{12}$H-*ras*.

### *Revertants of EJ/T24 bladder carcinoma cells*

EJ/T24 human bladder carcinoma cells contain a val$^{12}$H-*ras* transforming gene and no normal H-*ras* allele (Feinberg et al 1983). Three independent revertants, 1aCB, 8b Cl1 and 24d, were isolated from two separate subclones of EJ cells, EJ8 and EJ11. Fig. 3 shows that the revertants had a flat morphology. Unlike the parental cells, all three revertants were unable to proliferate under anchorage-independent conditions (Table 1). In order to examine whether the revertants still contained val$^{12}$H-*ras* transforming genes, DNA was prepared from each revertant and assayed for its transforming capacity in calcium phosphate transfection assays on NIH-3T3 cells. Table 2 shows that all three revertants retained val$^{12}$H-*ras* transforming genes and that the transforming efficiency of DNA from the revertants was similar to that of DNA from the parental cells.

The revertants from HT1080 cells that we have previously characterized were all sensitive to re-transformation by transforming *ras* proteins (Paterson et al 1987). We therefore investigated whether the revertants of EJ cells could

**TABLE 1 Properties of parental and revertant clones from EJ cells (val$^{12}$H-*ras*)**

| | | | | *Retransformation by p21ras* | | |
|---|---|---|---|---|---|---|
| | | | | *Transfection*[a] | | *Microinjection*[b] |
| *Parental* | *Colony-forming efficiency in agar* | *Revertant* | *Colony-forming efficiency in agar* | *% pSV2neo EJ-transformed colonies* | *% pSV2neo N-rasHT-transformed colonies* | |
| EJ8 | 80% | 1aCB | <0.1% | < 1% | < 1% | 0 |
| | | 24d | <0.1% | >95% | >95% | ND |
| EJ11 | 70% | 8b Cl1 | <0.1% | 1.7% | 1% | 0 |

[a] Revertant cell lines were transfected with 10 μg/plate plasmid DNA + 10 μg/plate normal mouse DNA as carrier using calcium phosphate (Wigler et al 1978). The val$^{12}$ c-H-*ras* oncogene was subcloned into pSV2neo to give pSV2neoEJ, and the gln$^{61}$ N-*ras* HT1080 oncogene was subcloned into pSV2neo to give pSV2neoN-*ras*HT (Paterson et al 1987). Transfected colonies were detected in G418 and the percentage of transformed morphology G418$^{R}$ colonies was scored.

[b] 50–100 cells were microinjected with val$^{12}$H-*ras* p21 protein at 3 mg/ml purified from an *E. coli* expression system (Trahey et al 1987). Microinjected cells were followed by time-lapse video microscopy to estimate the fraction that regained the transformed morphology of the parental EJ cells.

**TABLE 2 Transforming activity of genomic DNA isolated from parental EJ bladder carcinoma cells and their flat revertants in NIH-3T3 transfection assays**

| *Cell line* | *Transforming activity foci/μg DNA* |
|---|---|
| EJ8 | 0.5 |
| revertant 1aCB | 1.0 |
| revertant 24d | 0.6 |
| EJ11 | 0.7 |
| revertant 8b Cl1 | 1.0 |

be re-transformed by transfection with plasmids encoding the val$^{12}$H-*ras* transforming gene. Table 1 shows that revertant 24d, like the HT1080 revertants, could be re-transformed by transforming val$^{12}$H-*ras* protein and therefore retains sensitivity to transformation by oncogenically activated *ras* proteins. In contrast, revertants 1aCB and 8b Cl1 could not be re-transformed by transfection with val$^{12}$H-*ras* or lys$^{61}$N-*ras* plasmids.

### *The mechanism of reversion differs between the two ras-resistant revertants*

The revertants 8b Cl1 and 1aCB arose independently, because they were derived from separate subclones of EJ cells; it is therefore of interest to determine whether the mechanism of reversion is the same in each revertant. To investigate this question, somatic cell hybrids were made between each revertant and its parental clone and also between the two revertants. To facilitate the isolation of the cell hybrids, G418 resistance and hygromycin resistance as dominant selectable markers were introduced into the cell lines by plasmid transfection. Appropriate combinations of cells were then fused using PEG1500 and the resulting hybrids selected by culture in medium containing G418 and hygromycin.

Fig. 3e shows that hybrids between revertant 1aCB and parental EJ8 cells displayed the transformed morphology characteristic of EJ8 cells, whereas the predominant morphology of the hybrids between revertant 8b Cl1 and parental EJ11 cells was flat, like that of the revertant cells (Fig. 3f). Thus revertant 8b Cl1, unlike revertant 1aCB, is capable of suppressing the transformed phenotype of EJ bladder carcinoma cells. Fusion of the two revertants 1aCB and 8b Cl1 produced cells with the revertant morphology (Fig. 3h). As expected, fusions between the two parental lines, EJ8 and EJ11, led to cell hybrids with transformed morphology (Fig. 3g).

## Discussion

We have shown that flat revertants which have regained normal growth control, as assayed by anchorage-dependent growth, can be isolated from human carcinoma cells. In these revertants there is no possibility that rever-

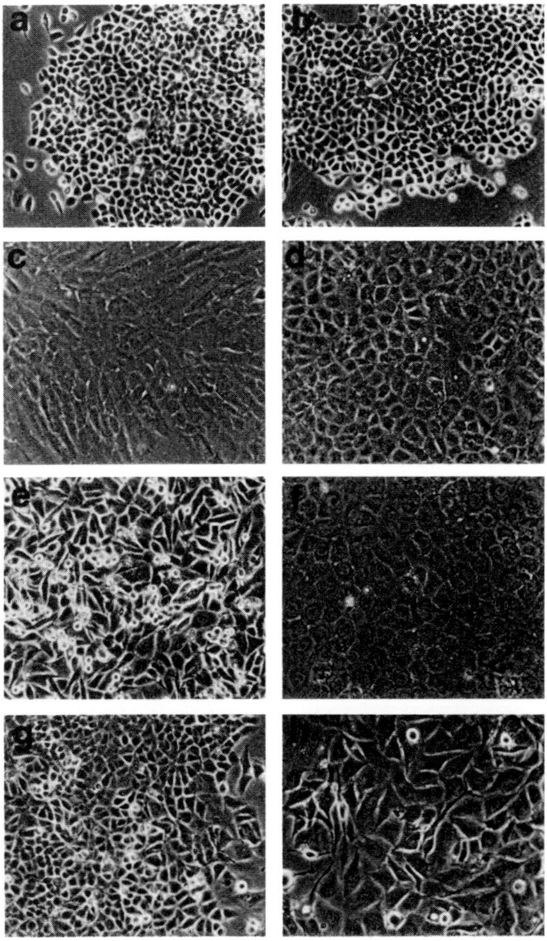

FIG. 3. Morphology of hybrids between parental EJ bladder carcinoma cells and revertant cell lines. (a) EJ8 hygro$^R$, (b) EJ11 neo$^R$, (c) revertant 1aCB neo$^R$, (d) revertant 8b Cl1 hygro$^R$, (e) hybrid 1aCB neo$^R$ × EJ8 hygro$^R$, (f) hybrid 8b Cl1 hygro$^R$ × EJ11 neo$^R$, (g) hybrid EJ8 hygro$^R$ × EJ11 neo$^R$, (h) hybrid 1aCB neo$^R$ × 8b Cl1 hygro$^R$. hygro$^R$, hygromycin-resistant; neo$^R$, neomycin-resistant.

sion arises by a suppressive action of the normal allele of the transforming *ras* gene, because the normal H-*ras* allele has been lost from EJ cells (Feinberg et al 1983). Furthermore, microinjecting mixtures of normal and oncogenically activated p21*ras* proteins, as we describe, shows it is unlikely that the normal p21*ras* product is capable of suppressing transformation by the oncogenically activated *ras* protein.

Two of the revertants of EJ cells, 8b Cl1 and 1aCB, are resistant to re-transformation by oncogenically activated p21*ras* proteins. The mechanism of reversion appears to be different in each case, because hybrids of revertant 8b Cl1 with EJ cells are non-transformed, whereas hybrids between revertant 1aCB and EJ cells are transformed. A simple genetic interpretation of these observations would be that revertant 8b Cl1 contains a dominant gene capable of suppressing *ras* transformation (cf Noda et al 1983), whereas revertant 1aCB contains a recessive defect. Such an interpretation is consistent with the non-transformed nature of hybrids between the two revertants.

These experiments therefore point to a multiplicity of mechanisms leading to reversion of *ras* transformed cells and should provide the basis for attempts to isolate the genes involved.

*Acknowledgements*

This work was supported by the Cancer Research Campaign and the Medical Research Council.

## References

Adari H, Lowy DR, Willumsen BM, Der CJ, McCormick F 1988 Guanosine triphosphatase activating protein (GAP) interacts with the p21*ras* effector binding domain. Science (Wash DC) 240:518–521

Bos JL 1988 The ras-gene family and human carcinogenesis. Mutat Res 195:255–271

Calés C, Hancock JF, Marshall CJ, Hall A 1988 The cytoplasmic protein GAP is implicated as the target for regulation by the *ras* gene product. Nature (Lond) 332:548–551

Capon D, Seeburg P, McGrath J et al 1983 Activation of Ki-ras 2 gene in human colon and lung carcinomas by two different point mutations. Nature (Lond) 304:507–513

Feinberg A, Vogelstein B, Droller M, Baylin S, Nelkin B 1983 Mutation affecting the 12th amino acid of the c-Ha-ras oncogene product occurs infrequently in human cancer. Science (Wash DC) 220:1175–1177

Fleischman LF, Chawala SB, Cantley L 1986 Ras-transformed cells: altered levels of phosphatidylinositol 4,-5-bisphosphate and catabolites. Science (Wash DC) 231:407–410

Hancock JF, Marshall CJ, McKay IA et al 1988 Mutant but not normal p21*ras* elevates inositol phospholipid breakdown in two different cell systems. Oncogene 3:187–193

Lacal J-C, Fleming TP, Warren BS, Blumberg PM, Aaronson SA 1987 Involvement of functional protein kinase C in the mitogenic response to the H-ras oncogene product. Mol Cell Biol 7:4146–4149

Marshall C, Sager R 1981 Genetic analysis of tumorigenesis: IX. Suppression of anchorage independence in hybrids between transformed hamster cell lines. Somatic Cell Genet 7:713–723

Morris JDH, Price B, Lloyd AC, Self AJ, Marshall CJ, Hall A 1988 Scrape loading of Swiss 3T3 cells with *ras* protein induces rapid activation of protein kinase C followed by DNA synthesis. Oncogene, in press

Mulcahy LS, Smith MR, Stacey DW 1985 Requirement for *ras* proto-oncogene function during serum-stimulated growth of NIH-3T3 cells. Nature (Lond) 313:241–243

Noda M, Selinger Z, Scolnick E, Bassin R 1983 Flat revertants isolated from Kirsten sarcoma virus-transformed cells are resistant to the action of specific oncogene. Proc Natl Acad Sci USA 80:5602–5606

Paterson H, Reeves B, Brown R et al 1987 Activated N-ras controls the transformed phenotype of HT1080 human fibrosarcoma cells. Cell 51:803–812

Ricketts MH, Levinson AD 1988 High level expression of c-Ha-*ras*-1 fails to fully transform Rat-1 cells. Mol Cell Biol 8:1460–1468

Trahey M, McCormick F 1987 A cytoplasmic protein stimulates normal N-*ras* p21 GTPase but does not affect oncogenic mutants. Science (Wash DC) 238:542–545

Trahey M, Milley RJ, Cole G et al 1987 Biochemical and biological properties of the human N-*ras* p21 protein. Mol Cell Biol 7:541–544

Wakelam MJO, Davies SA, Houslay MD, McKay IA, Marshall CJ, Hall A 1986 Normal p21N-*ras* couples bombesin and other growth factor receptors to inositol phosphate production. Nature (Lond) 323:173–176

Wigler M, Pellicer A, Silverstein S, Axel R 1978 Biochemical transfer of single copy eukaryotic genes using total cellular DNA as donor. Cell 14:725–731

Willumsen BM, Christensen A, Hubbert NL, Papageorge AG, Lowy DR 1984 The p21*ras* C-terminus is required for transformation and membrane association. Nature (Lond) 310:583–586

## DISCUSSION

*Stanbridge:* There are two classes of HT1080 revertants, one of which has no activated N-*ras* gene. One might imagine that therefore you would not get re-revertants of that particular class, is that the case?

*Marshall:* We have never obtained tumours from the HT1080 revertant, which has lost the N-*ras* transforming allele, when we inject it into animals.

*Stanbridge:* What about morphology?

*Marshall:* Its morphology is rather different to anything that it came from: the cells are rounded-up, in contrast to the other class of revertant that we got, the flat revertants, but they do show the inability to grow in anchorage-independent conditions, and therefore they require high concentrations of growth factors.

*Stanbridge:* If one introduces by transfection the active N-*ras* into those revertants, is there conversion to the tumorigenic phenotype?

*Marshall:* Yes.

*Klein:* How does one reconcile the fact that tumour progression in Dr Balmain's studies showed that the loss of the normal *ras*-carrying chromosome does convey some selective advantage on the cell, whereas you do not see any competition in the mutant/normal *ras* mixture experiments? Is it possible that the normal *ras* gene exerts a *trans*-acting influence on the expression of the mutant gene? You described a gene dosage change from two to one in the revertant, then back to two in the re-revertant. Why should that make such great difference in phenotype? It is a very small difference.

*Marshall:* The only rational explanation is to say that there is a very sharp threshold between normal growth control and transformation brought about by the *ras* genes in these cells. The situation with Allan's (Balmain) experiments is that whenever you see loss of the normal allele you also see duplication of the mutant transforming allele. So that could be a mechanism whereby you increase the copies of the mutant, but the cell does not want three copies of that chromosome so it loses the chromosome that carries the normal allele.

*Bodmer:* The fact that a cell is homozygous and has lost the normal allele doesn't mean that there is selection against the normal allele. It could be selection for increased gene dosage of the mutated allele and the mitotic recombinant would do exactly that.

*Marshall:* The argument I would make is selection for increased gene dosage of the transforming gene.

*Balmain:* There is evidence for a tumour suppressor type gene on chromosome 1 which also carries the N-*ras* gene. In our case, the Harvey-*ras* gene in the mouse is on chromosome 7, which is the equivalent of the short arm of chromosome 11 in humans. So there may be a putative suppressive gene on chromosome 7 in the mouse. In both situations it may not be a simple question of loss of the normal gene, there may also be loss of a linked suppressor locus.

*Marshall:* That is a very reasonable proposition. One way to address that is to isolate a much larger series of revertants and ask whether there are any which increase the number of copies of the chromosome which carries the normal allele and increase the dosage of that suppressor. We are considering that sort of experiment.

*Varmus:* Are you able to show that the 1 aCB cell can be re-transformed after fusion with a normal cell?

*Marshall:* I haven't done that experiment but it's a good one to do.

*Benedict:* In certain cases you are talking about reversion of tumorigenicity and other morphological changes, what particular things did you look at?

*Marshall:* In the HT1080 cells, we look at both morphological changes in tissue culture and tumorigenicity. With the EJT24 cells we've concentrated much more on the tissue culture aspects because in our hands the EJ cells sometimes give very variable results in nude mice.

*Klinger:* Chris, do you know anything about the ouabain sensitivity of your revertants, as compared to the parental cells?

*Marshall:* No.

*Wyke:* In the microinjection experiments, you had to use 208F because of its altered response to normal *ras* p21. Do you think that the lack of response by 208F to normal p21 compromises the conclusion of these microinjection experiments?

*Marshall:* I suspect that the Rat-1 derivative, 208F, is more like a normal fibroblast cell than NIH-3T3 cells. One way you could argue why 208F doesn't respond to overexpression of normal *ras* is that it is very resistant to the *ras* transforming signal, therefore it is more sensitive to things which down-regulate that signal.

*Stanbridge:* Chris, is there a single GTPase-activating protein that will enhance the activity of all three *ras* gene products or is there a different one for each *ras* gene?

*Marshall:* I don't know. Sigal (Merck, West Point, Pennsylvania) who has cloned and expressed a GTPase-activating protein (GAP) (personal communication) would have the answer to that.

*Weinberg:* I believe that the GAP activates the activity of at least two of them. I imagine it will be found to activate all three.

*Noble:* Chris, in the experiments that Anne Ridley, Hartmut Land, Hugh Paterson and I did with Schwann cells, the interpretation of our results was that function of *ras* within a Schwann cell was completely dependent on the presence or absence of a nuclear oncogene (Ridley et al 1988). If an oncogene was present, *ras* behaved as a transforming protein; in the absence of any nuclear oncogene, *ras* induced proliferation arrest. That makes me wonder whether revertants might represent effectors within the nuclear oncogene pathway rather than in the *ras* pathway.

*Marshall:* It is a very interesting possibility. The other feature of the Schwann cell experiments is that whether *ras* causes proliferation or inhibition is very dependent on the conditions the cells are in. In the presence of high concentrations of growth factors, *ras* inhibits growth; in low concentrations of growth factors, *ras* stimulates growth. It is almost as though the cell is weighing up the signals—if there is too much, growth factors and *ras*, the cells stop. If there is not enough signal coming from the growth factors, *ras* will cause proliferation.

*Adams:* This finding that *ras* alone often causes growth arrest raises the possibility that over-expression of certain kinds of *ras* genes might itself cause a revertant phenotype. Is there any evidence for that?

*Noda:* This question has been addressed by several groups and is still controversial. We are trying to repeat these experiments with normal Harvey-*ras* and N-*ras*.

*Marks:* Chris, does the depletion of protein kinase C by TPA cause changes in any of your hybrids?

*Marshall:* No, we haven't done those experiments. I don't know whether phorbol esters will completely down-regulate protein kinase C in these cells. For protein kinase C down-regulation experiments with *ras*, we used Swiss 3T3 cells which have been so well characterized by Henry Rozengurt and Peter Parker (Rodriguez-Pena & Rozengurt 1984).

*Noda:* We are trying to isolate genes which suppress the transformed phenotype associated with an activated *ras* oncogene. Our approach is to take transformed NIH-3T3 cells containing two copies of Kirsten murine sarcoma virus and transfect these with a cDNA expression library made from messenger RNA of normal human fibroblasts and then isolate the revertants.

First we screen the revertant by its morphology and test its biological activity in several different ways. Then we recover the cDNA from the interesting revertants and test the biological activities of these cDNAs. The parental cells

**TABLE 1** (*Noda*) **Patterns of oncogene suppression in cell hybrids between cells transformed with the oncogene indicated and revertants of NIH-3T3 cells transformed with Kirsten sarcoma virus then transfected with cDNA from normal human fibroblasts**

| *Revertant* | *ras* | *src* | *mos* | *raf* | *fos* |
|---|---|---|---|---|---|
| R16 | + | − | − | − | − |
| R31 | + | + | −? | − | − |
| R14/R37 | + | + | + | − | − |
| R12 | + | + | − | + | − |
| R29/40 | + | + | + | + | + |

have typical transformed morphology: attachment to the substratum is loose, they form colonies in soft agar and form tumours in nude mice. We isolated seven independent revertants. The morphology varies slightly between clones but they all have dramatically decreased tumorigenicity in nude mice, almost the same as NIH-3T3 cells. They do not proliferate in agar suspension culture or in medium with low concentrations of serum. We have evidence that in those revertants biologically active *ras* genes are present and making good amounts of the oncogene product, p21.

If we fuse those revertants with NIH-3T3 cells, they stay flat, therefore the revertant phenotype is dominant in somatic cell hybrids. We also found that the transformed phenotype was suppressed in hybrids between Kirsten sarcoma virus-transformed NIH-3T3 cells and each of the revertants. Then we took cells transformed by various other oncogenes and fused them with each revertant. By the patterns of specificity for the suppression of transformed phenotypes associated with these oncogenes, we could classify the revertants into five classes (Table 1). So our results are consistent with the idea that cDNA transfected into the cell is responsible for the revertant phenotype.

So far we have isolated the cDNAs from two of the revertants, the R16 and R40 revertants. Hitoshi Kitayama in our laboratory sequenced the cDNA named K*rev*-1 isolated from R16. We cannot find any homology between this sequence and any one in the data banks. But the amino acid sequence derived from the long open reading frame in this cDNA did show some homology with, to our surprise, *ras* proteins. Interestingly, the region in c-H-*ras* protein known to interact with GTP is very well conserved in the predicted amino acid sequence for K*rev*-1. The putative effector binding site, which is believed to be the domain that interacts with GAP, is also conserved.

Our current speculation is that either K*rev*-1 protein might interact with the same target as *ras* and interfere with its growth stimulatory signal or K*rev*-1 protein might represent components similar to *ras* in the pathway of signal transduction for negative growth regulation.

*Bodmer:* Is it possible that one way to get a revertant is through a mutation in one of the other *ras* genes? The *ras* genes have somewhat related functions, so you could envisage a mutant in another member of the family in some way competing with the first mutational activity.

*Noda:* Pierre Chardin and Armand Tavitian in France identified the same gene by cross-hybridization with a *Drosophila ras-3* probe, and the sequence was identical.

*Bodmer:* So is your clone a new member of the *ras* family?

*Noda:* Yes, but I think it is not a mutant, it is a wild-type gene.

*Noble:* That's very interesting. M. Nieto-Sampedro (1988) has looked at the ability of astrocytes to inhibit the effect of EGF. The inhibitory substance secreted by astrocytes has homology with the EGF receptor and binds EGF in the medium and thereby depletes it. Is this something we have all missed? Is there a common regulatory strategy involving a whole class of proteins that are endogenous inhibitors?

*Weinberg:* I would suggest that this new gene is not physiologically an inhibitor but represents a gene product which functions analogously to the *ras* proteins in normal signal transduction. By over-expression in the vectors that Dr Noda has created, it happens to bind all the available molecules of some target protein which is required by the oncogenic *ras* protein to transform cells. Therefore just thc over-expression is able to interdict the pathway that normally allows the cell to respond productively to the transforming effects of *ras* p21.

*Marshall:* There is an important feature of the K*rev*-1/*rap*-1 predicted gene product—at codon 61 it has a threonine residue that in the *ras* protein would be a transforming allele with reduced intrinsic GTPase activity. The prediction is that the K*rev*-1/*rap*1 protein in the cell is probably in the GTP form, whereas normal *ras* is probably most of the time in the GDP form. So K*rev*-1/*rap*-1 protein might hold the target for *ras* in some 'off position', then when *ras* gets converted to the GTP form, it can displace K*rev*-1/*rap*-1 protein and activate that target.

*Klein:* How is K*rev*-1 protein expressed in the revertants you have isolated previously?

*Noda:* We haven't tested this.

*Sporn:* Most cells that have been transformed by *ras* are insensitive to TGF-β growth regulation, even when the parental cell is sensitive. Have you looked at whether your revertants are regulated by TGF-β?

*Noda:* No; we should try those experiments.

*Stanbridge:* These revertants have this interesting pattern of resistance to a variety of other oncogenes. If it were a negative effect on *ras*, one might think that particular cDNA might not have a reverting effect on, for example, a *src*-transformed cell? Have you introduced the cDNA clone into a *src*-transformed cell?

*Noda:* We have only done cell fusion with NIH-3T3 cells transformed by *src*. We have not directly transfected this gene into *src*-transformed cells, but in the cell fusion experiments it did not suppress *src*, it seems to be specific for *ras*.

*Varmus:* Dennis Stacey has shown that an anti-*ras* antibody directed against normal p21 can interfere with transformation by *src*. Your results with R16 suggest that the R16 clone doesn't interfere with transformation by *src*, although it specifically interferes with transformation by an activated Kirsten-*ras* allele. Thus, it doesn't affect whatever normal p21 does to mediate *src* transformation. Have you tested R16 for interference with transformation by Harvey-*ras* or N-*ras*?

*Noda:* Yes, it suppressed transformation by Harvey-*ras* in cell fusion experiments. We haven't tested N-*ras*.

*Varmus:* Were those codon 12 mutants?

*Noda:* Yes, because we used only viral oncogenes in those experiments.

## References

Nieto-Sampedro M 1988 Astrocyte mitogen inhibitor related to epidermal growth factor receptor. Science (Wash DC) 240:1784–1786

Ridley AJ, Paterson HF, Noble M, Land H 1988 *ras*-mediated cell cycle arrest is altered by nuclear oncogenes to induce Schwann cell transformation. EMBO (Eur Mol Biol Organ) J 7:1635–1645

Rodriguez-Pena A, Rozengurt E 1984 Disappearance of $Ca^{2+}$-sensitive phospholipid-dependent protein kinase activity in phorbol ester-treated 3T3 cells. Biochem Biophys Res Commun 120:1053–1059

# A genetic basis for tumour suppression

Eric J. Stanbridge

*Department of Microbiology and Molecular Genetics, California College of Medicine, University of California, Irvine, California 92717, USA*

*Abstract.* The technique of somatic cell hybridization has established the phenomenon of tumour suppression and provided evidence for a genetic basis for suppression. Further refinements aimed at eventually identifying 'tumour suppressor' genes include the use of monochromosome transfer via microcell hybridization. The application of this technique to the study of tumour suppression in tumorigenic HeLa cell × fibroblast hybrids, Wilms' tumour, retinoblastoma and osteosarcoma cells is described. The issue of whether tumour suppression involves a direct effect on expression of activated oncogenes is discussed. Transformation of normal human cells in culture by activated cellular oncogenes is an extremely rare event. This may be due to a relatively greater genomic stability of human cells compared to rodent cells. We describe the use of a spontaneously immortalized human keratinocyte cell line, HaCaT, for studies of the effects of introduction of activated c-Ha-*ras* oncogene into these cells, with particular reference to tumorigenic conversion.

*1989 Genetic analysis of tumour suppression. Wiley, Chichester (Ciba Foundation Symposium 142) p 149–165*

The idea that chromosomal changes are in some way involved in the genesis of cancer has its origin in the studies of Boveri at the turn of the century. With no prior knowledge of the existence of genes or mutation he postulated that chromosomal abnormalities were the key cellular changes leading to the transition from a normal cell to a malignant one.

This remarkable foresight has been borne out in the succeeding years. A genetic basis for cancer is now well established and includes the discovery of oncogenes, whose activation or altered expression is commonly associated with the cancerous state, and more recently the putative 'tumour suppressor' genes. In this latter category, it seems that the loss or inactivation of tumour suppressor genes is a necessary condition for neoplastic growth, that is, they seem to be negative controlling elements.

Reports of the phenomenon of tumour suppression actually predate the discovery of oncogenes (Shih et al 1981) and stem from the pioneering somatic cell hybrid studies of Harris, Klein and their colleagues (Harris et al 1969, Harris 1988). The early groundwork for analysis of tumorigenesis by somatic cell genetics was laid by analysis of intraspecies rodent and interspecies human–rodent hybrid cells. The most informative studies, which analysed hybrid cells generated from the fusion of malignant and normal parental cells, reported suppression of tumorigenicity. Tumorigenic segregants were readily obtained from these non-tumorigenic hybrids and this re-expression of tumorigenicity was subsequently shown to be due to the loss of specific chromosomes (Evans et al 1982, Sager 1985). Unfortunately, a significant number of the somatic cell hybrids examined in these studies were relatively unstable and rapidly lost chromosomes. This chromosomal instability led to some controversy regarding the phenomenon of tumour suppression and to claims that malignant expression behaved as a dominant trait (Croce 1984).

The problem of chromosomal instability was overcome by using intraspecies human cell hybrids, which proved to be very stable in culture (Stanbridge et al 1981). A summary of our extensive analysis of human carcinoma (HeLa) × diploid fibroblast hybrid cells is presented in Table 1 (also see Stanbridge 1976, Stanbridge 1987). The phenomenon of tumour suppression was confirmed in this experimental system. Suppression of tumorigenicity has now been demonstrated in a variety of human malignant × normal cell fusions, allowing one to conclude that tumour suppression is a fairly general consequence of introducing genetic information from normal cells into cancer cells. Several other interesting features arose from these studies. The non-tumorigenic hybrid cells continued to behave as transformed cells in culture (a property shared with hybrid cell combinations from a variety of species), indicating that the transformed and tumorigenic phenotypes are under separate genetic control and supporting the idea that progression from a normal cell to a neoplastic one involves multiple genetic alterations. It should also be noted that with certain malignant × normal cell fusions, even the transformed phenotype is suppressed, resulting in hybrid cells that behave as normal cells which, in culture, eventually senesce and die (Pereira-Smith & Smith 1983).

Extensive cytogenetic and restriction fragment length polymorphism (RFLP) analyses indicated that re-expression of tumorigenicity in HeLa × fibroblast hybrids was associated with the loss of specific chromosomes (Stanbridge et al 1981, Klinger 1982), particularly chromosome 11 derived from the normal fibroblast (Srivatsan et al 1986, Kaelbling & Klinger 1986).

## Monochromosome transfer studies

The fact that a single chromosome seems to contain the genetic information required for control of the tumour-forming ability of HeLa represents a welcome decrease in complexity from that which would prevail if genes on multiple chromosomes were required in concert [a possibility that may yet prove to be correct (Klinger 1982)]. However, the conclusion that chromosome 11 is involved in control of tumorigenic expression of HeLa is based on indirect evidence only. A direct confirmation of this possibility is to transfer

**TABLE 1 Summary of the properties of intraspecies human cell hybrids (HeLa × diploid fibroblast)**

1. Fusion of tumorigenic HeLa and normal fibroblasts results in hybrid cells that are non-tumorigenic.
2. The non-tumorigenic hybrid cells behave as transformed cells in culture. Therefore, the transformed and tumorigenic phenotypes appear to be under separate control.
3. Rare tumorigenic segregants arise from the non-tumorigenic hybrid cell populations.
4. Re-expression of tumorigenicity is correlated with the loss of a single copy of the fibroblast chromosome 11 (determined by cytogenetic analysis and restriction fragment length polymorphism analysis).
5. Introduction of a fibroblast chromosome 11 into tumorigenic HeLa × fibroblast hybrids via microcell transfer results in suppression of tumorigenicity.
6. Suppression of tumorigenicity is associated with the induction of differentiation *in vivo*.

single chromosomes from normal cells into cancer cells and examine their effect on tumorigenicity.

The technique of microcell transfer allows one to test such an hypothesis (Fournier & Ruddle 1977, Saxon et al 1985). The key feature of this technique is that, as with whole cell fusions, the single chromosome transferred by microcell fusion is retained in succeeding cell generations as a complete chromosome and is not degraded, as occurs with metaphase chromosome transfer.

Using this technique it has been possible to show that transfer of a chromosome 11 (actually an X;11 translocation chromosome which contains the majority of the genetic information present on chromosome 11) derived from a normal human fibroblast into tumorigenic segregant HeLa × fibroblast hybrids restores the control of tumorigenic expression and results in non-tumorigenic hybrids (Saxon et al 1986). Rather surprisingly, it was found that transfer of the X;11 chromosome into HeLa cells also resulted in suppression of tumorigenicity. On the basis of the analysis of HeLa × fibroblast hybrids, one would expect two copies of fibroblast chromosome 11 to be necessary for complete tumour suppression. Recently, we have determined that this tumour suppression of HeLa/X;11 microcell hybrids is not complete and that tumours do form, albeit more slowly than with the HeLa parent. Experiments are in progress to determine the effect of introducing a second copy of a fibroblast chromosome 11 into the HeLa/X;11 microcell hybrids.

The loss of genetic information, as evidenced by specific chromosome deletions and loss of specific RFLP alleles, is associated with a significant number of human cancers (see Table 2). In one such case, retinoblastoma, the gene that is often altered or deleted in this cancer, the *Rb-1* gene, has been cloned (Friend et al 1986). A direct measure of the effect of restoring *Rb-1* gene function in retinoblastoma cells that lack this gene product, by

**TABLE 2 Human solid tumours with associated specific cytogenetic deletions and loss of RFLP alleles**

| *Type of cancer* | *Chromosome region involved* | *References* |
|---|---|---|
| Retinoblastoma | 13q14 | Cavenee et al (1983) |
| Wilms' nephroblastoma | 11p13 | Koufos et al (1984) |
| Familial renal cell carcinoma | 3p14 | Wang et al (1984) |
| Uveal melanoma | 2 | Mukai et al (1986) |
| Neuroblastoma | 1p32→pter | Brodeur et al (1981) |
| Acoustic neuroma | 22q | Seizinger et al (1986) |
| Sporadic renal cell carcinoma | 3p | Zbar et al (1987) |
| Small cell lung carcinoma | 3p13→p24 | Whang-Peng et al (1982) |
| Ductal breast carcinoma | 13q | Lundberg et al (1987) |
| | 11p | Ali et al (1987) |
| Colorectal carcinoma | 5q | Solomon et al (1987) |
| | 17p | Fearon et al (1987) |

introduction of the cloned gene via DNA transfection, has not yet been reported.

In the absence of the availability of such putative tumour suppressor genes, a direct test of the relevance of specific chromosome deletions associated with various human cancers is microcell transfer of the intact chromosome in question from a normal cell into the cancer cell. This has been accomplished with Wilms' tumour. Constitutional deletions of chromosome 11p13 appear to predispose to the disease and chromosome 11 abnormalities involving the same region are seen frequently in tumour tissue. Loss of RFLP alleles in this region have also been reported (Knudson 1986, Hansen & Cavenee 1987). Microcell transfer of the fibroblast X;11 chromosome (which contains the relevant 11p13 region) into a Wilms' tumour cell line resulted in suppression of tumorigenicity (Weissman et al 1987). Interestingly, these cells retained their transformed properties in culture. However, the non-tumorigenic Wilms'/X;11 microcell hybrids have a decreased ability to grow in suspension in soft agar and a dramatically decreased ability to grow in medium containing calf serum compared to the parental Wilms' tumour cells (C. Chang, personal communication). Cells which had lost the X;11 chromosome regained their ability to form tumours and to grow in medium containing calf serum.

Recently, attempts have been made to introduce a normal chromosome 13 (tagged with a neomycin resistance gene), which contains an intact *Rb-1* gene, into retinoblastoma cells (Y-79) and osteosarcoma cells (OSV) that have an intact *Rb-1* gene but appear to express a truncated *Rb-1* mRNA (Y. Fung, personal communication). No viable microcell hybrids have been recovered from the retinoblastoma Y-79 fusions. However, introduction of the chromosome 13 into OSV cells has been successful and again suppression of tumorigenicity is seen (Table 3) (Stanbridge et al 1989).

**TABLE 3 Tumorigenicity of human osteosarcoma cell line OSV and OSV/13 microcell hybrids**

| *Cell line* | *Number of cells inoculated/site* | *Number of tumours/Number sites inoculated* |
|---|---|---|
| OSV | $1 \times 10^7$ | 4/4 |
| MCH 311 Mass 1 | $1 \times 10^7$ | 0/6 |
| Mch 312 Mass 2 | $1 \times 10^7$ | 2/8 |
| Mch 312 Clone 1 | $3 \times 10^7$ | 0/5 |

It will be interesting to see if introduction of the cloned *Rb-1* gene into OSV cells has a similar tumour-suppressing effect. It is important to remember that introduction of a normal chromosome 13 into OSV cells represents transfer of a great deal of genetic information in addition to the *Rb-1* gene.

## Do tumour suppressor genes directly affect expression of activated oncogenes?

Evidence has accumulated that activated oncogenes play some, as yet unidentified, role in the genesis of cancer (Bishop 1983). Activation may occur as a consequence of mutation, translocation, amplification, inappropriate expression, *etcetera*. Given that putative tumour suppressor genes have such a dramatic effect on tumorigenicity, it is reasonable to expect that they may decrease or extinguish the level of activated oncogene expression. Examination of somatic cell hybrids, where the tumorigenic parental cell expresses an activated oncogene, has in most cases not borne out this prediction (Craig & Sager 1985, Geiser et al 1986).

In each case, suppression of tumorigenicity occurs with continued expression of the activated oncogene at a level commensurate with that seen in the parental cancer cell. In one instance even the fusion of two cancer cells containing different activated *ras* oncogenes resulted in a hybrid that was non-tumorigenic and which continued to express both activated oncogenes (Geiser et al 1989). Flat, non-tumorigenic revertants of MSV-transformed rodent cells also continue to express the viral *src* oncogene (Noda et al 1983). There are one or two reports where expression of the oncogene is down regulated in non-tumorigenic hybrids. Wyke and colleagues have reported that non-tumorigenic RSV-transformed × normal rodent fibroblasts express negligible amounts of the viral *src* oncogene (Green & Wyke 1985). It has also been reported that flat non-tumorigenic revertants of the human fibrosarcoma cell line HT1080, which contains an activated allele of N-*ras*, express correspondingly lower levels of the mutant p21 protein than do their tumorigenic counterparts (Paterson et al 1987). The situation is clearly complicated and will probably vary from one type of cancer cell to the next.

## Malignant transformation of human cells with activated *ras* oncogenes

Many of our interpretations of the role of oncogenes in cancer have been derived from studies of the introduction of viral oncogenes or human cellular oncogenes into cultured rodent cells resulting in malignant transformation (Bishop 1983, Varmus 1984). However, whereas rodent cells are transformed readily following introduction of activated oncogenes and also undergo spontaneous transformation rather frequently, spontaneous immortalization and neoplastic transformation are extremely rare events in human cell culture systems. Introduction of *myc* and *ras* oncogenes into cultured normal diploid fibroblasts has little effect on their normal phenotype beyond a transient increase in the ability to grow in soft agar, even though the oncogenes are expressed (Sager et al 1983, Hurlin et al 1987, E. Stanbridge, unpublished observations). These observations indicate that these cellular oncogenes do not cause malignant transformation of human diploid fibroblasts.

The majority of human cancers are carcinomas and therefore of epithelial origin. There have been two reports of successful transformation of human epithelial cells, both using viral *ras* oncogenes for the transfection experiments (Rhim et al 1985, Yoakum et al 1985). In one case, immortalization of the epithelial cells with SV40 before infection with the Kirsten sarcoma virus was a prerequisite (Rhim et al 1985). In the other case, transfection of bronchial epithelial cells with the v-Ha-*ras* oncogene was followed by a long latency period until an immortal cell line developed and a similar period of several months passed before the cells were capable of forming tumours in nude mice (Yoakum et al 1985). Thus, it is entirely possible that further genetic alterations were necessary to obtain malignant bronchial epithelial cells, independent of *ras* gene expression. Indeed, this method of transforming human bronchial epithelial cells is not readily reproducible (C. Harris, personal communication).

It seems, therefore, that transformation of cultured normal human diploid cells (whether fibroblast or epithelial) with activated cellular *ras* oncogenes is an extremely rare event compared to the relative ease with which rodent cells can be malignantly transformed. This may be due, in part, to the proposed greater genomic stability of human cells (Sager 1984). Human cells that have been immortalized with viral oncogenes may have acquired greater genomic instability and therefore become more susceptible to *ras* oncogene-induced malignant transformation.

This possibility has been examined using a recently isolated, spontaneously immortalized human keratinocyte cell line, designated HaCaT (Boukamp et al 1988). These cells are aneuploid but, despite exhibiting phenotypic traits of transformation in culture, are non-tumorigenic in nude mice. HaCaT cells were transfected with a plasmid construct containing the neomycin resistance gene (for selection purposes) and the activated EJ c-Ha-*ras* oncogene (Parada et al 1982). Colonies were selected on the basis of resistance to geneticin, an

**TABLE 4** ***In vivo*** **growth behaviour of HaCaT and HaCaT-*ras* transfectants in nude athymic mice**

| Cell line | | Regressing nodules | Persisting benign cysts | Squamous cell carcinomas | Mutant EJ-ras p21 expression |
|---|---|---|---|---|---|
| Parent | | | | | |
| HaCaT | | 12[a] | 0 | 0 | – |
| HaCaT-*ras* clones | | | | | |
| Class I | clone I-6 | 3 | 0 | 0 | – |
| | A-4 | 5 | 0 | 0 | – |
| Class II | clone I-5 | 3 | 3 | 0 | – |
| | A-5 | 1 | 3 | 0 | – |
| | I-7 | 7 | 3 | 0 | + |
| Class III | clone B-6 | 2 | 0 | 2 | + |
| | II-3 | 1 | 1 | 21 | + |
| | A-6 | 1 | 0 | 8 | + |
| | I-8 | 0 | 0 | 5 | ± |
| | B-8 | 0 | 1 | 6 | – |

[a] Numbers of mice in each category. Each mouse was inoculated subcutaneously with $5 \times 10^6$ cells. Adapted from Boukamp et al (1989).

analogue of neomycin, and examined for altered phenotypic traits in culture, mutant p21 protein expression and tumorigenic potential. There was no alteration in growth behaviour *in vitro* of any of the HaCaT-*ras* clones examined. When tested for tumorigenic potential the clones fell into three classes (Table 4). Class I clones were non-tumorigenic and, like the HaCaT parental cells, formed small regressing nodules in nude mice. When class II clones were inoculated into nude mice, the nodules in a proportion of the injected animals significantly and consistently enlarged. Histological examination of these nodules revealed that they were benign cysts with extensive areas of keratinization within the cysts, accompanied by peripheral areas of hyperplastic and dysplastic epidermal tissue architecture. These cysts, following a period of enlargement over several months, persisted without further progression or gradually regressed. For class III clones, a significant proportion of mice inoculated developed highly differentiated invasive squamous cell carcinomas. When the clones were examined for mutant *ras* p21 protein expression by Western blot analysis it was found that class I clones did not express any mutant form of the protein. In classes II and III mutant p21 protein was found. Most of the clones which formed invasive squamous cell carcinomas expressed the mutant form of p21 but two clones had barely detectable levels. Of three class II clones tested, two had no measurable mutant p21 and one had a level equivalent to that of class III clones. Thus, although there is a trend toward expression of EJ *ras* p21 and altered growth behaviour of HaCaT-ras cells *in vivo*, exceptions occur in each group. We believe a reasonable interpretation of these results is that expression of EJ *ras* p21 protein in HaCaT cells confers a limited proliferative advantage on the cells which may result in hyperplastic or dysplastic growth *in vivo* (Boukamp et al 1989). Further genetic or epigenetic alterations are presumably necessary for the cells to acquire the ability to form invasive squamous cell carcinomas. This interpretation is consistent with experimental carcinogenesis models of mouse skin tumours, where initiation–promotion is followed by the appearance of benign papillomas which contain an activated c-Ha-*ras* oncogene (Balmain 1985, Strickland et al 1988). The further genetic alterations could, of course, include the inactivation of tumour suppressor gene(s).

## Perspectives

A growing body of evidence, initially from somatic cell hybrid studies but now from a variety of experimental sources, has established the phenomenon of tumour suppression and a genetic basis for its action. The negative control of malignant expression would seem, at least in certain instances, to be *trans*-acting. This *trans*-acting negative regulation generally does not appear to influence directly the expression of conventional activated oncogenes found in the cancer cells that are converted to a non-tumorigenic phenotype. Furthermore, the role that these activated oncogenes play in the cancer induction process is questioned by the studies outlined here. Although much work still needs to be done to ascertain the role of oncogene activation in cancer, it does seem that, at least in human cancers, activation of *ras* oncogenes may predispose a cell to further genetic changes that eventually result in malignant behaviour but expression of the activated oncogene alone is insufficient to endow the cell with tumour-forming potential.

Clearly, the tumour suppressor genes do not affect expression of the dominantly expressed conventional oncogenes. If their activity is *trans*-acting and regulatory in nature, it leads to the intriguing possibility that a class of oncogenes exist which are regulated in *trans* and will only be expressed if the *trans*-acting regulatory gene is inactivated or the target regulatory region of the putative oncogene is disrupted in some way. A candidate oncogene product has been identified in HeLa cells which behaves in this fashion (Der & Stanbridge 1981, Sutherland et al 1986). This class of oncogenes will obviously be more difficult to identify than those which are dominantly expressed.

As George Klein has stated (Klein 1987), we are in the emerging era of tumour suppressor genes. It will be necessary to characterize these genes at the molecular level and identify their functions in order to ascertain their true role in the control of malignant transformation. This should now be possible.

*Acknowledgements*

The author's studies were supported by grants from the National Institutes of Health and Council for Tobacco Research, USA. I am indebted to Sabine Eckgold for preparation of this manuscript.

## References

Ali IU, Lidereau R, Theillet C, Callahan R 1987 Reduction to homozygosity of genes on chromosome 11 in human breast neoplasia. Science (Wash DC) 238:185–188

Balmain A 1985 Transforming *ras* oncogenes and multistage carcinogenesis. Br J Cancer 5:1–7

Bishop JM 1983 Cellular oncogenes and retroviruses. Annu Rev Biochem 52:301–354

Boukamp P, Petrussevska RT, Breitkreutz D, Hornung J, Markham A, Fusenig NE 1988 Normal keratinization in a spontaneously immortalized aneuploid human keratinocyte cell line. J Cell Biol 106:761–771

Boukamp P, Stanbridge EJ, Foo DY et al 1989 Lack of correlation between activated c-Ha-ras oncogene expression and malignant transformation or altered differentiation potential of the human HaCaT keratinocyte cell line. submitted for publication

Brodeur GM, Green AA, Hayes FA et al 1981 Cytogenetic features of human neuroblastomas and cell lines. Cancer Res 41:4678–4686

Cavenec WK, Dryja TP, Phillips RA et al 1983 Expression of recessive alleles by chromosomal mechanisms in retinoblastoma. Nature (Lond) 305:779–784

Craig RW, Sager R 1985 Suppression of tumorigenicity in hybrids of normal and oncogene-transformed CHEF cells. Proc Natl Acad Sci USA 82:2062–2066

Croce CM 1984 Gene regulation in the expression of malignancy. Cancer Surv 3:287–298

Der CJ, Stanbridge EJ 1981 A tumor-specific membrane phosphoprotein marker in human cell hybrids. Cell 26:429–438

Evans EP, Burtenshaw MD, Brown BB, Hennion R, Harris H 1982 The analysis of malignancy by cell fusion. IX. Reexamination and clarification of the cytogenetic problem. J Cell Sci 56:113–130

Fearon ER, Hamilton SR, Vogelstein B 1987 Clonal analysis of human colorectal tumors. Science (Wash DC) 238:193–197

Fournier REK, Ruddle FH 1977 Microcell-mediated transfer of murine chromosomes into mouse, Chinese hamster and human somatic cells. Proc Natl Acad Sci USA 74:319–323

Friend SH, Bernards R, Rogelj S et al 1986 A human DNA segment with properties of the gene that predisposes to retinoblastoma and osteosarcoma. Nature (Lond) 323:643–646

Geiser AG, Der CJ, Marshall CJ, Stanbridge EJ 1986 Suppression of tumorigenicity with continued expression of the c-Ha-ras oncogene in EJ bladder carcinoma-human fibroblast hybrid cells. Proc Natl Acad Sci USA 83:5209–5213

Geiser AG, Anderson MJ, Stanbridge EJ 1989 Suppression of tumorigenicity in human cell hybrids derived from cell lines expressing different activated *ras* oncogenes. Cancer Res, in press

Green AR, Wyke JA 1985 Anti-oncogenes. A subset of regulatory genes involved in carcinogenesis? Lancet 2 (8453) 475–477

Hansen MF, Cavenee WK 1987 Genetics of cancer predisposition. Cancer Res 47:5518–5527

Harris H, Miller OJ, Klein G, Worst P, Tachibana T 1969 Suppression of malignancy by cell fusion. Nature (Lond) 223:363–368

Harris H 1988 The analysis of malignancy by cell fusion: the position in 1988. Cancer Res 48:3302–3306

Hurlin PJ, Fry DG, Maher VM, McCormick JJ 1987 Morphological transformation, focus formation and anchorage independence induced in diploid human fibroblasts by expression of a transfected H-ras oncogene. Cancer Res 47:5752–5757

Kaelbling M, Klinger HP 1986 Suppression of tumorigenicity in somatic cell hybrids. III. Cosegregation of human chromosome 11 of a normal cell and suppression of tumorigenicity in intraspecies hybrids of normal diploid × malignant cells. Cytogenet Cell Genet 41:65–70

Klein G 1987 The approaching era of the tumor suppressor genes. Science (Wash DC) 238:1539–1544

Klinger HP 1982 Suppression of tumorigenicity. Cytogenet Cell Genet 32:68–84

Knudson AG Jr 1986 Genetics of human cancer. Annu Rev Genet 20:231–251

Koufos A, Hansen MF, Lampkin BC et al 1984 Loss of alleles on human chromosome 11 during genesis of Wilms' tumour. Nature (Lond) 309:170–172

Lundberg C, Skoog L, Cavenee WK, Nordenskold M 1987 Loss of heterozygosity in human ductal breast tumors indicates a recessive mutation on chromosome 13. Proc Natl Acad Sci USA 84:2372–2376

Mukai S, Dryja TP 1986 Loss of alleles at polymorphic loci on chromosome 2 in uveal melanoma. Cancer Genet Cytogenet 22:45–53

Noda M, Selinger Z, Scolnick EM, Bassin RH 1983 Flat revertants isolated from Kirsten sarcoma virus-transformed cells are resistant to the action of specific oncogenes. Proc Natl Acad Sci USA 80:5602–5606

Parada LF, Tabin CJ, Shih C, Weinberg RA 1982 Human EJ bladder carcinoma oncogene is homologue of Harvey sarcoma virus ras gene. Nature (Lond) 297:474–478

Paterson H, Reeves B, Brown R et al 1987 Activated N-ras controls the transformed phenotype of HT1080 human fibrosarcoma cells. Cell 51:803–812

Pereira-Smith OM, Smith JR 1983 Evidence for the recessive nature of cellular immortality. Science (Wash DC) 221:964–966

Rhim JS, Jay G, Arnstein P et al 1985 Neoplastic transformation of human epidermal keratinocytes by AD12-SV40 and Kirsten sarcoma viruses. Science (Wash DC) 227:1250–1252

Sager R 1984 Genetic instability, suppression and human cancer. Cancer Surv 3:321–334

Sager R 1985 Genetic suppression of tumor formation. Adv Cancer Res 44:43–68

Sager R, Tanaka K, Lau CC et al 1983 Resistance of human cells to tumorigenesis induced by cloned transforming genes. Proc Natl Acad Sci USA 80:7601–7605

Saxon PJ, Srivatsan ES, Leipzig GV et al 1985 Selective transfer of individual human chromosomes to recipient cells. Mol Cell Biol 5:140–146

Saxon PJ, Srivatsan ES, Stanbridge EJ 1986 Introduction of human chromosome 11 via microcell transfer controls tumorigenic expression of HeLa cells. EMBO (Eur Mol Biol Organ) J 5:3461–3466

Seizinger BR, Martuza RL, Gusella JF 1986 Loss of genes on chromosome 22 in tumourigenesis of human acoustic neuroma. Nature (Lond) 322:644–647

Shih C, Padhy LC, Murray M, Weinberg RA 1981 Transforming genes of carcinomas and neuroblastomas introduced into mouse fibroblasts. Nature (Lond) 290:261–264

Solomon E, Voss R, Hall V et al 1987 Chromosome 5 allele loss in human colorectal carcinomas. Nature (Lond) 328:616–619

Srivatsan ES, Benedict WF, Stanbridge EJ 1986 Implication of chromosome 11 in the suppression of neoplastic expression in human cell hybrids. Cancer Res 46:6174–6179

Stanbridge EJ 1976 Suppression of malignancy in human cells. Nature (Lond) 260:17–20

Stanbridge EJ 1987 Genetic regulation of tumorigenic expression in somatic cell hybrids. Adv Viral Oncol 6:83–101

Stanbridge EJ, Flandermeyer RR, Daniels DW, Nelson-Rees WA 1981 Specific chromosome loss associated with the expression of tumorigenicity in human cell hybrids. Somatic Cell Mol Genet 7:699–712

Stanbridge EJ, Weissman BE, Fung YT, Benedict WF 1989, in preparation

Strickland JE, Greenhalgh DA, Kocevachyla A et al 1988 Development of murine epidermal cell lines with contain an activated Ha-ras oncogene and form papillomas in skin grafts on athymic nude mouse hosts. Cancer Res 48:165–169

Sutherland DR, Bicknell DC, Downward J et al 1986 Structural and functional features of a cell surface phosphoglycoprotein associated with tumorigenic phenotype in human fibroblast × HeLa cell hybrids. J Biol Chem 261:2418–2424

Wang N, Perkins KL 1984 Involvement of band 3p14 in t (3;8) hereditary renal carcinoma. Cancer Genet Cytogenet 11:479–481

Weissman BE, Saxon PJ, Pasquale SR et al 1987 Introduction of a normal chromosome 11 into a Wilms' tumor cell line controls its tumorigenic expression. Science (Wash DC) 236:175–180

Whang-Peng J, Bunn PA, Kaoshan CS et al 1982 A nonrandom chromosomal abnormality, del 3p (14–23) in human small cell lung cancer (SCLC). Cancer Genet Cytogenet 6:119–134

Varmus HE 1984 The molecular genetics of cellular oncogenes. Annu Rev Genet 18:553–612

Yoakum GH, Lechner JF, Gabrielson EW et al 1985 Transformation of human bronchial epithelial cells transfected by Harvey *ras* oncogene. Science (Wash DC) 227:1174–1179

Zbar B, Brauch H, Talmadge C, Linehan M 1987 Loss of alleles of loci on the short arm of chromosome 3 in renal cell carcinoma. Nature (Lond) 327:721–724

## DISCUSSION

*Sporn:* Do you have some thoughts on why, in the chromosome 13 transfer studies, there was a difference between the retinoblastoma cells and the osteosarcoma cells?

*Stanbridge:* It could be trivial, i.e. that we have failed to introduce the chromosome 13 into the retinoblastoma cells. The alternative is that the introduction of chromosome 13 results in a non-proliferating or differentiating cell.

*Noble:* Which is what you would predict from the biology of the retina. A fundamental difference between the osteogenic cells and neural cells is that the osteogenic cells don't go into a terminally non-dividing state.

*Stanbridge:* That is reasonable but we need to do the experiment to prove it. The key experiment will be to introduce a chromosome other than 13 into the retinoblastoma cells; if they continue to proliferate, then I would feel more comfortable with that interpretation.

*van der Eb:* Are you able to get G418-resistant retinoblastoma cells when you introduce the chromosome 13?

*Stanbridge:* Not at all, we obtain no viable cells.

*van der Eb:* Could it be that you simply can't get G418 resistance in these tumour cells?

*Stanbridge:* That is a possibility.

*Benedict:* The problem is that retinoblastoma cells like to clump when they grow in suspension. We have tried various tricks to make them grow as single cells, and then done microcell fusions. However, they are not easy fusions to carry out. There could be also some trivial reason why we are not getting results with these retinoblastomas.

*Bodmer:* Can't you just do it by transfection?

*Benedict:* Nobody has been able to do that yet.

*Stanbridge:* Bob, you have attempted some transfections using constructs containing the neomycin resistance gene. Have you obtained any neomycin-resistant retinoblastoma cells?

*Weinberg:* Yes, but it's very difficult. They don't transfect well and when they do take up DNA they don't clone very well at low density. When you apply the selection, the few colonies that have been transfected don't grow, they don't cross-feed one another adequately.

*Stanbridge:* We did reconstruction experiments, plating the retinoblastoma cells at low densities, and used conditions that gave us a reasonable rate of recovery of proliferating cells. However, a low transfection rate would certainly make it difficult to obtain G418-resistant clones.

Dr Boukamp in Heidelberg has taken those highly differentiated squamous cell carcinomas from the nude mice, re-established cells from them in culture, and reintroduced them into nude mice. They again formed highly differentiated invasive squamous cell carcinomas.

*Tarin:* You mentioned that Wilms' cells which have received a chromosome had become less able to grow in serum.

*Stanbridge:* Yes, in medium containing calf serum. There wasn't a significant difference in their growth rate in fetal calf serum but there is a dramatic difference in new-born calf serum. That suggests there may be differences in growth factor requirements.

*Weinberg:* Gian Paolo Dotto did a series of experiments while he was in my laboratory. He took mouse skin cells, keratinocytes, and infected them with a *ras*-containing virus. He then put them in Fusenig chambers on the backs of mice. These chambers allow keratinocytes and dermal fibroblasts to be cultured *in situ* in an area from which the skin has been removed, resulting in the reconstitution of skin within the chamber.

If the population of cells put into the chamber consists purely of *ras*-containing keratinocytes, then one gets a rapidly expanding, undifferentiated squamous cell carcinoma. If, on the other hand, one mixes those cells with a three or four-fold excess of normal dermal fibroblasts and puts them into the identical environment, nothing happens. One gets at best only a nodule, which eventually regresses and disappears. This suggests that tumorigenicity of a cell is not only dependent on its own genotype and the oncogenes it bears, but is enormously influenced by its normal neighbours.

*Stanbridge:* That is true, even under *in vitro* conditions in your own experiments and those of others, and it is also very much the case in an *in vivo* microenvironment.

*Marks:* It would be interesting to determine whether you can modulate the phenotype of the tumours in animals which have been given one of the polar inducer compounds. In a number of cell lines that we've looked at, if there is even a low level of differentiation, as seems to be the situation with the tumours you study, one can markedly increase the proportion of cells that differentiate by changing the environment, specifically, by exposure to a polar compound, such as HMBA. Such inducer-mediated differentiation is associated with a suppression of tumorigenicity (Marks et al 1978).

*Stanbridge:* Yes, it's a very worthwhile experiment.

*Marks:* In your experiments examining possible effects of *ras* in EJ x fibroblast hybrids, is it possible that without any change in the level of *ras* expression, there is an altered structure of the *ras* protein which correlates with tumorigenicity?

*Stanbridge:* I doubt it, because we always go from the non-tumorigenic to the tumorigenic form. Secondly, one can take the DNA of non-tumorigenic cells, transfect 3T3 cells with that DNA, get the equivalent level of transformation frequency and the gene which is doing that transformation is the original activated *ras*.

*Adams:* You described a possible correlation between tumorigenicity and the amount of *ras* expression. Were you looking after the tumour had developed or simply looking at cells that would go on to give those results?

*Stanbridge:* We have re-established cells from the tumours into cell culture. In these tumour cell reconstituted cell lines there is an equivalent level of expression of p21, irrespective of whether the cells are tumorigenic or not.

*Noble:* In your last experiment with the HaCT/*ras* cells, if instead of putting the epithelial cells on a collagen layer, you put them directly onto muscle cells, or if in culture you simply place them on a muscle cell feeder layer, what are the consequences?

*Stanbridge:* I have no information using a muscle cell feeder layer. The HaCaT/*ras* cells have been placed directly onto the muscle fascia by Dr Boukamp. The results are a little difficult to interpret, because mouse mesenchymal cells tend to come up over the surface of the implanted cells. There are definitely features of differentiation into stratified epithelial tissue. However, the Class III clones that form the invasive squamous cell carcinomas rapidly show features of local invasion. The differences in degree of differentiation seen with the three classes of HaCaT/*ras* clones may be due to differences in response to the tissue microenvironment.

*Klinger:* As you are aware, there is a problem with the microcell results in that only one copy of chromosome 11 [t(X;11) in this case] from a non-tumorigenic cell was able to suppress the D98 cell. Data from your group and our own have shown that two copies of a normal chromosome 11 were needed to suppress the tumorigenic phenotype of D98 HeLa cells (Stanbridge et al 1981, Kaelbling & Klinger 1986, Klinger & Kaelbling 1986).

*Stanbridge:* It is a very interesting situation. The whole cell hybrid results with HeLa x fibroblast hybrids showed that when two copies of chromosome 11 from the fibroblast and two copies of chromosome 11 from HeLa are present, the cell is non-tumorigenic. In the tumorigenic segregants, Professor Klinger and I have shown that the loss of a single copy of the normal chromosome 11 is sufficient to produce a tumorigenic phenotype. These HeLa x fibroblast tumorigenic segregants were the cells into which we introduced a further normal chromosome 11 [t(X;11)]. These cells now had a non-tumorigenic phenotype. That suggests that if one took HeLa alone and introduced a single copy of a normal chromosome 11, that cell should still be tumorigenic, and that it would require the introduction of two copies of chromosome 11 to render it non-tumorigenic. The surprising result was that we appeared to get non-tumorigenic HeLa cells after the introduction of one copy of a normal chromosome 11. In fact, subsequent studies showed that tumour suppression is seen but is not complete. We get what we would rather call partial suppression, that is a slower growth, but we are getting tumours. We weren't able in those early studies to put in two copies of chromosome 11 because you need a different selectable marker on each copy. We now have that capacity and those experiments are underway.

*Harris:* We have a situation parallel to this in the case of chromosome 4 of the mouse. When one makes hybrids between a range of mouse tumour cells and normal mouse fibroblasts, there is strong selective pressure *in vitro* and *in vivo* against the normal fibroblast chromosome 4. By the use of translocations we have obtained some idea of where the operative locus is situated. In these mouse hybrid cells, where you have both copies of the normal fibroblast chromosome 4 you don't get any tumours. You start getting a few tumours when one copy of the normal chromosome is eliminated. But if you look at those tumours, you find there is further selective pressure against the normal chromosome 4. In many of the tumours, you will find that the cells have eliminated both copies of the normal chromosome 4, but have doubled up one of the chromosomes 4 from the malignant cell. If you do a second set transplant or serial transplants, you eliminate all the chromosomes 4 derived from the normal fibroblast.

In a normal cell, we can assume that we have two functional copies of the allele that we are interested in, but there has been very little information about how many functional alleles there are in the tumour cell that we are fusing the normal cell with. The increasing evidence for functional homozygosity or loss of heterozygosity in tumour cells suggests that in the tumour cell that we are studying we may not be starting with two functional alleles.

*Klinger:* There is one further important point concerning this question of a dosage effect in suppression of tumorigenicity of D98 cells. We have shown that there has been loss of chromosome 11 heterozygosity in D98 (HeLa) cells, at least for a substantial segment of the short arm, where the restriction fragment length polymorphism (RFLP) markers used are located (Kaelbling et al 1986). Therefore, analogous to the situation in Wilms' tumour and retinoblastoma, there could be complete loss of a functional suppressor in D98 cells. If this is true, then one copy of a normal chromosome 11 might be expected to be enough to get suppression, but, as noted earlier, with the exception of Eric's single chromosome transfer experiments, two copies of the normal 11 were always necessary for suppression, suggesting that the suppressor gene product must reach a threshold level, i.e. the amount made from both alleles, for complete and stable suppression.

In your experiment with the G401 cell line, you didn't have very conclusive evidence that it was derived from a Wilms' tumour.

*Stanbridge:* We have no convincing data at all, only that the American Type Culture Collection identified it as such. If it is not a Wilms' tumour, it is a tumour cell line that when we introduce a chromosome 11 into it, becomes non-tumorigenic.

*Klinger:* We have failed to obtain cytogenetic or biochemical evidence for G401 being derived from a Wilms' tumour. I believe Web Cavenee has similar but more extensive results. I realize this does not prove that G401 is not of Wilms' tumour origin, but without evidence that it is, I fail to see what new information we have obtained from your experiments.

*Cavenee:* We've looked at expression patterns of about 50 different genes in various regions of developing kidney, adult kidney, in a few Wilms' tumours and in the cell line G401. The one that's different in that picture is G401, when compared to either kidneys or Wilms' tumours; although there is variation amongst Wilms' tumours, the total pattern is different from that in G401.

*Stanbridge:* We would like to take a *bona fide* Wilms' tumour cell line and repeat these experiments, but unfortunately they are not readily available.

*Klinger:* We have a proven Wilms' tumour cell line. The cells do not grow in nude mice, like most Wilms' cells I believe, unlike G401 which is highly tumorigenic in these mice. Kornel Schuebel in my laboratory fused these Wilms' cells with D98 cells to determine if the suppressor on the chromosome 11 that is assumed to be defective in the Wilms' cell is the same one that suppresses the D98 cell in hybrids with normal cells. If the same locus were involved, the Wilms' cell should not be able to suppress the D98 cell. The study is far from complete, in particular we do not yet have the critical results of the chromosome analyses, but Kornel has got some suppressed hybrids, suggesting that the Wilms' cell has the locus that suppresses the D98 cell, i.e. that the

Wilms' locus is not the same as the one that suppresses D98, although both are presumably on chromosome 11.

*Stanbridge:* I hesitate to put too much emphasis on tumour × tumour hybridization.

*Klinger:* I believe intertumour hybrids are good for asking the type of question I just discussed, that is for determining if suppressors in different tumour types belong to the same complementation group. In addition, such studies may provide considerable information on the interaction of tumorigenicity-promoting genes. Some of the hybrids in the experiment just described with Wilms' cells are much more tumorigenic than the parental D98 cells, which are of intermediate tumorigenicity ($5 \times 10^6$ D98 cells have to be injected to produce tumours in 100% of the mice within three weeks). If I may take a dangerous step and predict results, then I predict that in these hybrids there is loss of the chromosomes 11 of the Wilms' cell carrying the putative suppressors, so that tumorigenicity-promoting factors of both tumour parents are expressed. This is supported by the observation that the hybrid cell is tumorigenically enhanced ($5 \times 10^6$ cells produce tumours in 7–10 days). We obtained similar results when we fused D98 with a very weakly tumorigenic lymphoma cell (minimal tumour-producing dose of $10^7$ cells). Thus, depending on the chromosomes retained in such hybrids, you probably can get complementation of either suppressors or tumorigenicity-promoting factors. I believe these systems could also be of value in determining the relationship and interaction between different tumorigenicity-promoter and suppressor genes.

*Stanbridge:* The results of the D98 × Wilms' cell fusion agree with some studies that Bernard Weissman at the University of Southern California has done (personal communication). He has shown that Hela × G401 (Wilms' tumour) hybrids are non tumorigenic. Thus, it is not a problem as far as I am concerned, because there is no reason to suspect that the gene on chromosome 11 which appears to be controlling tumorigenic expression of HeLa is the same gene as the Wilms' tumour locus. We actually have some preliminary evidence from Dr Eri Srivatsan at the University of Southern California that that may not be the case.

## References

Kaelbling M, Klinger HP 1986 Suppression of tumorigenicity in somatic cell hybrids. III. Cosegregation of human chromosome 11 of a normal cell and suppression of tumorigenicity in intraspecies hybrids of normal diploid x malignant cells. Cytogenet Cell Genet 42:65–70

Kaelbling M, Roginski RS, Klinger HP 1986 DNA polymorphisms indicate loss of heterozygosity for chromosome 11 of D98AH2 cells. Cytogenet Cell Genet 41:240–244

Klinger HP, Kaelbling M 1986 Suppression of tumorigenicity in somatic cell hybrids. IV. Chromosomes of normal human cells associated with suppression of tumorigenicity in hybrids with D98AH2 carcinoma cells. Cytogenet Cell Genet 42:255–235

Marks PA, Reuben R, Epner E et al 1978 Induction of murine erythroleukemia cells to differentiate: a model for the detection of new anti-tumor drugs. Antibiot Chemother 23:33–41

Stanbridge EJ, Flandermeyer RR, Daniels DW, Nelson-Rees WA 1981 Specific chromosome loss associated with the expression of tumorigenicity in human cell hybrids. Somatic Cell Genet 7:699–712

# Molecular basis for the regulation of cell fate by the *lethal (2) giant larvae* tumour suppressor gene of *Drosophila melanogaster*

Bernard M. Mechler, Istvan Török, Martina Schmidt, Martin Opper, Anne Kuhn, Reinhard Merz and Ursula Protin

*Institute of Genetics, Johannes Gutenberg University, Saarstrasse 21, D-6500 Mainz, Federal Republic of Germany*

*Abstract.* Tumour suppressor genes act as recessive determinants of cancer. Their function is required for normal cell growth and differentiation during development. When both alleles of these developmental genes are inactivated, cell growth becomes unrestricted. In *Drosphila*, a series of genes have been identified which when mutated produce tissue-specific tumours. Of these the *lethal(2)giant larvae (l(2)gl)* gene is the best studied. Homozygous *l(2)gl* mutations cause the development of malignant tumours in the brain and the imaginal discs. Genomic DNA from the *l(2)gl* locus has been cloned, introduced back into *l(2)gl* mutant animals by P-element-mediated transformation and shown to restore normal development. The nucleotide sequence of the *1(2)gl* gene (13.1 kb) has been determined, as well as the sequences of the two classes of transcripts. These transcripts encode two polypeptides of 127 kDa and 78 kDa, respectively. Both proteins have been immunologically identified. Analyses of the spatial distribution of both *l(2)gl* transcripts and proteins revealed that during early embryogenesis the *l(2)gl* gene is uniformly expressed in all cells and tissues. In late embryos, the *l(2)gl* expression becomes gradually restricted to tissues presenting no morphological or neoplastic alteration in the mutant animals. Further mosaic experiments pointed out that the critical period for the establishment of tumorigenesis is limited to early embryogenesis at a time when the *l(2)gl* expression is most intense in all cells.

*1989 Genetic analysis of tumour suppression. Wiley, Chichester (Ciba Foundation Symposium 142) p 166–180*

Genetic analysis of *Drosophila* has revealed that cancer can be a genetic disorder and results from the inheritance of recessive mutant alleles at loci that have the characteristics of tumour suppressors. In *D. melanogaster* some 20 such loci have been identified (Gateff 1978). Homozygous recessive mutations in about two thirds of these loci interrupt the differentiation of adult primordial cell types and lead to an uncontrolled, invasive cell growth resulting in the premature death of the mutant animals (Gateff & Mechler 1988). These animals develop malignant tumours in the optic adult neuroblasts, the imaginal disc cells, or the blood cells. Mutations in six other loci cause benign gonial cell neoplasms resulting in the sterility of the mutant animals without impairing their development.

Similar tumour suppressor genes have been associated with genetic predisposition to tumour formation in familial cases of human cancer. One of the most thoroughly studied examples is the childhood disease retinoblastoma, for which it has been postulated that tumorous development of retina cells may proceed from homozygous loss of a locus that normally suppresses the tumour phenotype (Knudson 1971). The recent isolation of a gene with properties corresponding to the retinoblastoma gene (Dryja et al 1986, Friend et al 1986, Lee et al 1987, Fung et al 1987) has provided experimental support for the existence of tumour suppressor genes in human (see Knudson, this volume). Furthermore, suppression of tumorigenicity has been experimentally demonstrated by fusing normal and malignant cells. The resulting hybrid cells, which maintain a relatively complete chromosome complement, have lost the tumorigenic phenotype. However, in rare occasions, tumorous expression may reappear when specific chromosomes are eliminated (Harris et al 1969, Stanbridge 1976).

In *D. melanogaster*, the best characterized of these loci is the *lethal(2)giant larvae (l(2)gl)* gene. Homozygous mutations of the *l(2)gl* locus are responsible for the malignant transformation of two distinct tissues — the neuroblasts of the presumptive adult optic centres in the larval brain and the imaginal disc cells (Gateff & Schneiderman 1969, 1974). In homozygous *l(2)gl*-deficient animals, these malignancies first become visible in the third larval instar prior to the metamorphosis of the larvae. The tumorous growth produces a complex syndrome characterized by bloating of the larvae, underdevelopment of the ring gland and atrophy of the fat bodies. However, the underdevelopment of other tissues like the gonads and the imaginal cells of the salivary glands occurs before the outgrowth of the tumours and can thus be directly attributed to the absence of *l(2)gl* gene activity in these tissues. (For review see Hadorn 1955.)

The first *l(2)gl* mutant was discovered by Bridges in 1933 (Bridges & Brehme 1944) and has been mapped to position 21A-B at the extreme left end of the second chromosome (Lewis 1945). Subsequently, numerous spontaneous mutant alleles have been isolated from wild populations at several locations in the Soviet Union (Golubovsky 1978, 1980) and in California (Green & Shepherd 1979). Examination of naturally occurring variants revealed that about 1% of the second chromosomes were consistently mutated at the *l(2)gl* locus. The high incidence of *l(2)gl* mutations from geographically widespread populations and the high frequency of *l(2)gl* deletions following ethylmethanesulphonate mutagenesis of laboratory strains suggest that *l(2)gl* is contained within an unstable chromosomal region, presumably linked to telomeric variation.

## Molecular cloning and identification of the *l(2)gl* gene

The *l(2)gl* gene was isolated during a search for pole cell-specific transcripts (Mechler et al 1985). Pole cells are the first cells formed in the *Drosophila* embryo and become the precursors of the germline. They represent one of the rare examples of stem cells that can be identified and followed from the time of their formation up to their terminal differentiation as gametes. A *Drosophila* genomic DNA clone, α-8, found among a series of clones exhibiting pole cell specificity, was assigned by *in situ* hybridization to the 21A band at the left tip of the second chromosome. This clone was used to initiate a chromosomal walk which covered the entire *l(2)gl* region. The limits of the *l(2)gl* gene were determined by mapping 20 distinct chromosomal rearrangements within the *l(2)gl* locus and by relating the structural disruptions to one of the transcription units contained within the chromosomal walk. This procedure identified a transcription unit of about 13 kb which encodes two classes of overlapping transcripts of about 4.5 and 6 kb, respectively. In all the examined *l(2)gl* mutant alleles this transcription unit was structurally altered, either deleted or disrupted by the insertion of mobile repetitive elements (Mechler et al 1985).

Positive identification of the biologically active *l(2)gl* sequence was obtained by reintegrating a 13.1 kb DNA fragment containing the putative *l(2)gl* gene into the genome of *l(2)gl*-deficient animals by P-element-mediated germline transformation and showing that this sequence can fully rescue the development of the animals which normally would have succumbed to brain and imaginal disc neoplasia (Opper et al 1987).

## Molecular structure of the *l(2)gl* gene, transcripts and proteins

The sequence of 13.1 kb of DNA including and flanking the *l(2)gl* gene has been determined, as well as the sequence of several cDNAs representing both size classes of transcripts. From these data, the positions of the intron–exon junctions have been determined and the amino acid sequences of two *l(2)gl* polypeptide gene products have been predicted (Jacob et al 1987).

Structurally, the *l(2)gl* gene can be divided into two almost equal parts as shown in Fig. 1. The distal half contains most of the *l(2)gl* coding sequence whereas the proximal half consists of two direct repeats (α and β), each about 2.8 kb. Sequence comparison of the repeats indicates that they share a high degree (96%) of homology.

As revealed by sequence analysis of the 5′ end of seven distinct cDNAs, initiation of transcription occurs at the same location in both repeats. The

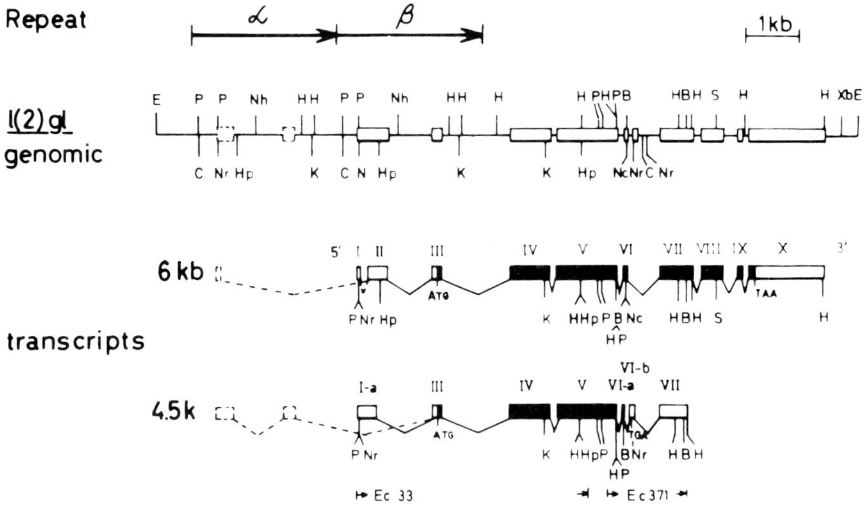


FIG. 1. Transcriptional organization of the *l(2)gl* gene. The proximal moiety of the *l(2)gl* gene consists of two direct repeats α and β, each of 2.8 kb. Both repeats contain a functional promoter. The structure of the two size classes of *l(2)gl* transcripts is indicated under the genomic DNA map. Exons are drawn as boxes and the coding sequences are shown in black. The structure of the 6kb class of transcripts is derived from cDNA clone Ec173, that of the 4.5 kb class of transcripts is based on the sequence of cDNA clones Ec33 and Ec371 (Jacob et al 1987). Sequence comparison of cDNA clones has revealed two differences in the exon splicing. Sequence variations found in the 5′ region of the transcripts do not affect the structure of the N-terminal domain of the *l(2)gl* proteins and can be attributed to the presence of duplicated promoters in the α and β repeats. When transcription starts from the α repeat the exons are indicated by dashed boxes[ ]. Both promoter regions are located about 0.5 kb downstream from the beginning of each repeat. By contrast, the sequence divergence identified in the middle of the gene modifies the coding capacities. The 6kb class of transcripts contains a large open reading frame encoding a protein of 1161 amino acids, p127, whereas a divergence characterizing the 4.5kb class of transcripts occurs in exon VI and produces a shorter protein of 708 amino acids, p78. Both proteins have been immunologically identified on Western blots of electrophoretically separated proteins from *Drosophila* embryos. Restriction sites: B, *Bam*HI; C, *Cla*I; E, *Eco*RI; H, *Hin*dIII; Hp, *Hpa*I; K, *Kpn*I; Nc, *Nco*I; Nh, *Nhe*I; Nr, *Nru*I; P, *Pvu*I; S, *Stu*I; X, *Xba*I.

presence of duplicated promoter regions leads to the synthesis of two classes of *l(2)gl* primary transcripts which differ in the length of the 5′ region. Depending on the splicing pattern, one to three exons from each repeat become assembled at the 5′ end of the final *l(2)gl* transcripts (Fig. 1). These variations do not affect the *l(2)gl* protein structure. However, a functional *l(2)gl* transcript should possess the exon III derived from the β repeat (exon $III_{\beta}$) because this exon contains the initiation codon for the large open reading

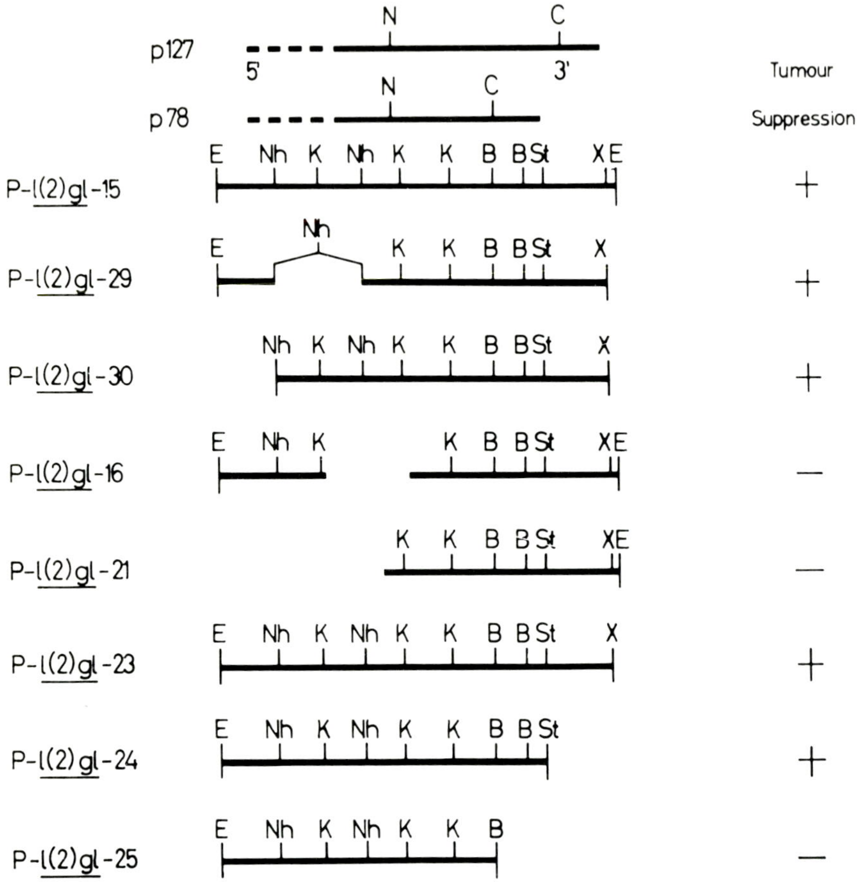


FIG. 2. Physical maps of the *l(2)gl* transposons used for transformation of *Drosophila* embryos. The *l(2)gl* structural gene (13.1 kb *Eco*RI fragment) has been inserted into the polylinker of the pUChs-neo-P-element vector and transformation has been conducted according to the protocol described in Opper et al 1987. Above the *l(2)gl* genomic sequence is shown the extent of the two transcripts (including the introns) that encode the p127 and p78 proteins. The dashed lines represent the transcripts initiated in the first repeat. The positions of the initiation and termination codons are indicated by N and C, respectively. On the right are shown the functional capacities of the P-*l(2)gl* transposons to complement *l(2)gl* deficiencies in tumour suppression. Restriction sites: B, *Bam*HI; E, *Eco*RI; K, *Kpn*I; Nh, *Nhe*I; St, *Stu*I; X, *Xba*I.

frame. When transcription is initiated in the upstream α repeat, a duplicated exon III (exon $III_{\alpha}$), also containing an initiation codon, becomes assembled in front of the exon $III_{\beta}$. If translation begins at the exon $III_{\alpha}$ initiation codon, only a short polypeptide of 24 amino acids is produced because the initiation codon of exon $III_\alpha$ is not in frame with the downstream large open reading frame. Furthermore, there is a change in the nucleotide preceding the ATG sequence of exon $III_\alpha$ which may strongly reduce the translation efficiency of this codon so that translation is preferentially initiated at the exon $III_\beta$ initiation codon.

In both repeats, sequence analysis of several cDNAs has revealed that the major transcription start site is located about 500 nucleotides downstream from the beginning of the repeat. Thus the upstream *cis*-acting regulatory elements of the *l(2)gl* gene should be contained within this 500 bp segment.

As shown in Fig. 2, P-mediated gene transfer experiments have demonstrated that the promoter region of both repeats is fully able to govern the expression of the *l(2)gl* gene (M. Schmidt, M. Opper, B. Mechler, unpublished work 1988). In each promoter, S1 protection and primer extension analyses have revealed the presence of a major transcription initiation site, as well as a series of secondary sites dispersed in the immediate vicinity of this major site (I. Török, B. Mechler, unpublished work 1988). The occurrence of multiple initiation sites suggests that the RNA polymerase is not able to align precisely on a defined sequence within the *l(2)gl* promoter but uses scattered, weakly defined TATA-related sequences to initiate *l(2)gl* transcription. However, the use of scattered start sites seems not to affect the level and regulation of *l(2)gl* transcription.

cDNA sequence analysis has also revealed the occurrence of alternative splicing in the central coding region of the *l(2)gl* gene. Through this mechanism, the *l(2)gl* gene can produce two partially overlapping proteins characterized by different C termini (Jacob et al 1987).

The amino acid sequence of the *l(2)gl* polypeptide encoded by the 6 kb transcript has been deduced from the sequence of the Ec173 cDNA, which corresponds to a complete copy of one of the transcripts of the 6 kb size class. This transcript would produce a polypeptide of 1161 amino acids with a relative mass of 126.9. This protein is therefore designated p127. The structure of the other *l(2)gl* polypeptide can be predicted from the modification of the nucleotide sequence found in the Ec371 cDNA. This predicts a polypeptide of 708 amino acids with a relative mass of 78.1, designated as p78, which is identical to the N-moiety of p127, differing only by the last amino acid, a phenylalanine residue. On the basis of the sequencing data, the *l(2)gl* polypeptide products can be organized in two domains; a 'core' domain containing 707 amino acids common to both p78 and p127 proteins and a 'tail' domain corresponding to the additional 454 residues present in the large p127 protein. Using antibodies that specifically recognize a portion of the N-terminal domain common to both *l(2)gl* polypeptides, the 'core domain', we have identified these two polypeptides with the expected sizes of about 80kDa and 130kDa, respectively (R. Merz, B. Mechler, unpublished work 1988).

The information gained from the structural analysis of the *l(2)gl* transcripts indicates that RNA splicing regulates the production of two distinct size classes of transcripts and consequently of *l(2)gl* polypeptides. Results from both developmental analysis of *l(2)gl* gene expression and reverse genetic experiments further support the existence of at least two *l(2)gl* components with distinct functional properties.

## The *l(2)gl* tumour suppressor domain

Reverse genetic experiments with transposons containing *l(2)gl* gene segments truncated at their 3′ extremity have allowed us to delineate more precisely a region carrying the *l(2)gl* tumour suppressor function (Jacob et al 1987). As shown in Fig. 1c, resection of a relatively large 3′ portion of genomic sequences, as exemplified in the p-*l(2)gl*-24 transposon, did not affect the tumour suppressor function of the *l(2)gl* gene, although a 3′ terminal segment of 2607 nucleotides has been deleted. This segment contains the 1803 terminal nucleotides of the 6 kb transcript encoding the 141 C-terminal amino acid residues. However, in the remaining *l(2)gl* DNA fragment the sequence corresponding to the 4.5 kb transcript is not altered. An additional resection which deletes sequence common to both classes of *l(2)gl* transcripts, as in the case of p-*l(2)gl*-25, completely abolishes the *l(2)gl* activity. These results may be interpreted in two ways: 1) that the truncation only shortens the *l(2)gl* gene without affecting the stability of the resected transcript and its encoded polypeptide, or 2) that this truncation specifically destabilizes the 6 kb class of transcripts and therefore prevents the expression of the p127 protein. In the latter case, the tumour suppressor function can be associated with the p78 protein; in the former, the rescue from lethality produced by the p-*l(2)gl*-24 transposon would indicate that the C-terminal portion of the p127 protein is dispensable.

## *l(2)gl* gene activity

The developmental profile of *l(2)gl* activity, as determined by Northern blot analysis, has revealed that the *l(2)gl* gene appears to be active mostly during embryogenesis and during the late third larval instar (Mechler et al 1985). During these two periods of *l(2)gl* activity the two size classes of transcripts are differentially expressed. The 4.5 kb transcripts are most abundant during early embryogenesis, whereas the 6 kb class of transcripts is mostly present during mid- and late-embryogenesis, as well as during the second period of *l(2)gl* activity in late third instar larvae.

The pattern of embryonic transcription coincides with *l(2)gl* protein expression. Western blot analysis of developmental embryonic stage proteins using antibodies that recognize the core domain of the *l(2)gl* proteins has revealed that the p78 protein is predominantly expressed in 2 to 10 hour-old embryos, whereas the p127 protein is most abundant in embryos aged 8 to 24 hours after egg deposition (R. Merz, B. Mechler, unpublished work 1988).

These two phases of *l(2)gl* expression correspond to the two major periods of cell proliferation during *Drosophila* development. On the basis of this correlation, it is tempting to attribute to the *l(2)gl* gene products a direct repression of cell proliferation or an indirect repression by inducing cell differentiation. In this latter case, the arrest of cell proliferation would result from the limited mitotic potential of differentiating cells. However, this hypothesis does not take into consideration the fact that the *l(2)gl*-deficient embryos are perfectly able to complete their embryogenesis. No morphological or histological differences can be noticed in *l(2)gl*-deficient embryos, although the *l(2)gl* embryonic cells have already acquired tumorigenic potential, as shown by transplantation of *l(2)gl*-deficient embryonic cells into the abdomen of adult hosts (Gateff & Schneiderman 1974). The tumorigenic commitment of embryonic cells suggests that the critical period for the establishment of the tumorigenesis takes place during the early embryonic development, when intense *l(2)gl* gene activity occurs, as revealed by our previous molecular analysis.

It is therefore important to examine the regions of *l(2)gl* expression in wild-type embryos to investigate possible correlations between the spatio-temporal expression of *l(2)gl* transcripts and proteins and the fate of these tissues in *l(2)gl*-deficient animals.

## Spatio-temporal pattern of *l(2)gl* expression

The availability of molecular probes for the *l(2)gl* transcripts and antibodies for the *l(2)gl* proteins allows the examination of the spatial limits of *l(2)gl* expression during embryogenesis.

Hybridization of *l(2)gl* probes to tissue sections of *Drosophila* embryos revealed the spatial and temporal pattern of *l(2)gl* transcription shown in Fig. 3. In this experiment, the *l(2)gl* probe was a 9 kb DNA fragment that hybridized to both classes of *l(2)gl* transcripts.

*l(2)gl* transcripts are first detected at the syncytial blastoderm stage over all peripheral nuclei. At the cellular blastoderm stage (Fig. 3A-B), the hybridization signals become stronger and are found predominantly over the cytoplasm in all cells. Then the *l(2)gl* expression remains uniform and relatively intense over all embryonic cells during the gastrulation and germ band extension stages (Fig. 3C-D) until about 8.5 hours of embryonic development.

At the time of the dorsal closure and the closure of the midgut epithelium (about 10 hours of development), the first sign of regional expression appears. Intense hybridization signals are still detected throughout the embryo but the midgut and the salivary gland epithelia present particularly strong signals. Examination of late embryonic stages shows that this regional

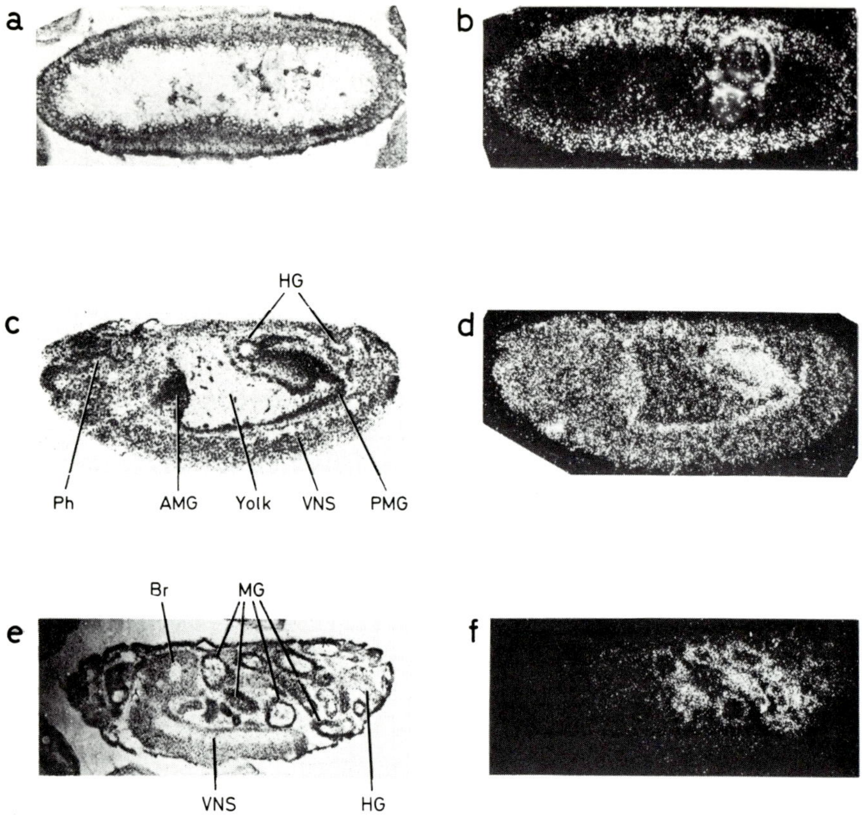


FIG. 3. Localization of *l(2)gl* transcripts in wild-type embryos. All sections are sagittal and oriented such that anterior is to the left and dorsal up. Localization of *l(2)gl* transcripts in (a, b) 2.5-hour, (c, d) 10.5-hour and (e, f) 18 to 20-hour embryos. b, d and f represent dark field photomicrographs of the same sections as in a, c and e, respectively. AMG, anterior midgut rudiment; Br, Brain; HG, hindgut; MG, midgut; Ph, pharynx; PMG, posterior midgut rudiment; VNS, ventral nervous system. From the formation of the syncytial blastoderm until germ band retraction, *l(2)gl* expression is uniformly observed in all embryonic cells and then becomes gradually restricted to the midgut and the salivary glands. Together with the results of the mosaic analysis and the damages observed in *l(2)gl* mutant animals, the results indicate that *l(2)gl* exerts pleiotropic functions which are dependent on the cellular context in which *l(2)gl* is expressed. Thus, the *l(2)gl* gene products probably interact with products of other genes which establish similar developmental programmes that govern important steps in terminal differentiation of larval and imaginal cells.

*l(2)gl* expression persists over the midgut and salivary gland epithelia, whereas the expression in other tissues has completely disappeared (Fig. 3E-F).

The analysis of the distribution of the *l(2)gl* proteins by immunocytochemical staining of tissue sections using anti-*l(2)gl* antibodies is underway.

Preliminary results indicate that the expression of *l(2)gl* proteins follows generally the pattern of *l(2)gl* transcription (R. Merz, B. Mechler, unpublished work 1988). During the early embryonic stages, from the cellular blastoderm until the germ band extension, a uniform staining is observed over all embryonic cells. In later embryonic stages, the distribution of the *l(2)gl* proteins exhibits regional differences in a pattern identical to that of expression of the *l(2)gl* transcripts. The distribution of the *l(2)gl* proteins becomes restricted to the midgut, the salivary glands and also to the axon projections forming the neuropile of the central nervous system. The accumulation of *l(2)gl* proteins in these axon projections indicates that *l(2)gl* protein turnover and translational efficiency may be different in the neuroblasts than in the other embryonic tissues.

A striking difference is also observed at the cellular level during embryonic development. In early embryos, the *l(2)gl* proteins are essentially localized in the cytoplasm, whereas at later stages the *l(2)gl* proteins become associated with the plasma membranes. Furthermore, examination of the midgut epithelium, which forms a monolayer of polyhedral cells, reveals that the *l(2)gl* proteins are uniquely associated with the plasma membrane facing contiguous cells and are completely absent from the basal and apical membranes of the epithelial cells.

The dual cytoplasmic and membrane distribution may reflect the successive expression of the two *l(2)gl* proteins. It is conceivable that the cellular localization of the p78 protein is cytoplasmic, whereas the p127 protein exhibits a plasma membrane association. At the molecular level the membrane association of the p127 protein is not yet understood: analysis of the amino acid sequence has not revealed a consensus hydrophobic N-terminal signal peptide indicating a membrane translocation, or a highly hydrophobic region that could form a transmembrane domain. However, different extracellular motifs, such as 13 Asn-X-Ser/Thr potential glycosylation sites and an Arg-Gly-Asp potential cell recognition tripeptide (Ruoslahti & Pierschbacher 1986), favour cell surface localization of p127. With respect to the presumed importance of the p78 protein in tumour suppression, the availability of antibodies recognizing specifically the p78 protein would greatly facilitate the analysis of this protein during early embryogenesis when the major molecular decision towards a normal developmental pathway is taking place.

## Critical period for the establishment of *l(2)gl* tumorigenesis

Recent analyses of *l(2)gl* mosaic animals indicate that the tumorous growth occurs in clones of cells which have lost the *l(2)gl* gene in the preblastoderm syncytial embryos, whereas clones produced at later embryonic stages do not display the tumour phenotype but are unable to complete differentiation (U. Protin, B. Mechler, unpublished work 1988). Finally, when the clones are produced during the larval stages, the *l(2)gl*-deficient cells show a nearly normal or normal development.

Providing that the *l(2)gl* activity is cell autonomous, these results show that the critical period for the establishment of *l(2)gl* tumorigenesis takes place during early embryogenesis, when the p78 protein is most intensively expressed.

### *l(2)gl* and the regulation of development

The molecular and developmental analysis of the *l(2)gl* gene has provided insight into both the biochemical nature of its products and its mode of action during development. The results indicate that *l(2)gl* plays a general role in development and contributes on a long term basis to the regulation of specific cell fates and the pattern of differentiation.

Analysis of patterns of expression of both *l(2)gl* transcripts and proteins revealed no direct correspondence with the phenotype of *l(2)gl*-deficient animals. Such a lack of correspondence was unexpected, because the simplest hypothesis regarding the role of *l(2)gl* in tumorigenesis predicted that the *l(2)gl* activity would be limited to the epidermal and neural precursors whose progeny cells are tumorigenic in the absence of *l(2)gl*. However, the results from the analysis of *l(2)gl* spatial expression during early embryogenesis and the mosaic analysis indicate that *l(2)gl* plays a much wider pleiotropic role in development, which would explain the pattern of developmental alterations observed in *l(2)gl*-deficient animals.

The neoplastic growth of neuroblasts and imaginal disc cells is certainly the most dramatic characteristic of the *l(2)gl* phenotype but it is only the ultimate feature among several pathological symptoms. Previous studies by Hadorn's group in the 1940s have shown that the absence of *l(2)gl* activity produces a complex pattern of damage affecting different cellular systems at different developmental stages. In particular, abnormalities in growth and differentiation have been noticed in the development of the male and female germlines, the salivary gland and the ring gland (Hadorn 1955).

All the *l(2)gl* developmental alterations result essentially from the absence of embryonic gene activity. This is in contrast to the developmental defects induced by mutations in pattern formation genes whose normal expression also occurs during early embryogenesis (for review, see Gehring 1987). In *l(2)gl* mutant embryos and larvae, morphogenesis proceeds normally but the cell fates of different adult anlagen are perturbed.

Our molecular and developmental analyses clearly indicate that the *l(2)gl* gene has different functions in several developmental programmes which are simultaneously established during embryogenesis and govern important steps in the terminal differentiation of larval and imaginal cell lineages. These different *l(2)gl* functions depend on the cellular context in which *l(2)gl* is expressed. Consequently, the *l(2)gl* products should interact with other gene products involved in similar programmes of development. Further genetic and biochemical studies will reveal the nature of these genes and gene products interacting with *l(2)gl*, and the phenotypes resulting from the absence of cooperation of these genes with *l(2)gl*. These studies will define the contribution of these genes in the establishment of tumorigenesis.

*Acknowledgements*

The authors wish to thank Elisabeth Gateff for continued interest and support, Dr. Dennis Strand for editorial assistance and Mrs H. Papadakis for secretarial help in preparing this manuscript. The studies reported here were supported by grants from the Deutsche Forschungsgemeinschaft SFB 302 — projects B20 and Me 800/1–2.

**References**

Bridges CB, Brehme KF 1944 The mutations of *Drosophila melanogaster*. Carnegie Institution of Washington Publication No 552

Dryja TP, Rapaport JM, Joyce JM, Petersen RA 1986 Molecular detection of deletions involving band q14 of chromosome 13 in retinoblastomas. Proc Natl Acad Sci USA 83:7391–7394

Friend SH, Bernards R, Rogelj S et al 1986 A human DNA segment with properties of the gene that predisposes to retinoblastoma or osteosarcoma. Nature (Lond) 323:643–646

Fung YKT, Murphee AL, T'Ang A, Qian J, Hinrichs SH, Benedict W 1987 Structural evidence for the authenticity of the human retinoblastoma gene. Science (Wash DC) 236:1657–1661

Gateff E 1978 Malignant neoplasms of genetic origin in *Drosophila melanogaster*. Science (Wash DC) 200:1448–1459

Gateff E, Mechler BM 1988 Tumor suppressor genes of *Drosophila melanogaster*. CRC Press, Boca Raton, Florida, in press

Gateff E, Schneiderman HA 1969 Neoplasms in mutant and cultured wild-type tissues of *Drosophila*. Natl Cancer Inst Monogr 31:365–397

Gateff E, Schneiderman HA 1974 Developmental capacities of benign and malignant neoplasms of *Drosophila*. Wilhelm Roux's Arch Dev Biol 176:23–65

Gehring WJ 1987 Homeo boxes in the study of development. Science (Wash DC) 236:1245–1252

Golubovsky MD 1978 The *lethal giant larvae* — the most frequent second chromosome lethal in natural populations of *Drosophila melanogaster*. Dros Inf Serv 53:179

Golubovsky MD 1980 Mutational processes and microevolution. Genetica (The Hague) 52/53:139–149

Green MM, Shepherd SHY 1979 Genetic instability in *Drosophila melanogaster*: the induction of specific chromosome 2 deletions by *mr* elements. Genetics 92:823–832

Hadorn E 1955 Lethalfaktoren. Thieme Verlag, Stuttgart

Harris H, Miller OJ, Klein G, Worst P, Tachibana T 1969 Suppression of malignancy by cell fusion. Nature (Lond) 223:363–368

Jacob L, Opper M, Metzroth B, Phannavong B, Mechler BM 1987 Structure of the *l(2)gl* gene of *Drosophila* and delimitation of its tumor suppressor domain. Cell 50:215–225

Knudson AG 1971 Mutation and cancer: statistical study of retinoblastoma. Proc Natl Acad Sci USA 68:820–823

Knudson AG Jr 1989 Epidemiology of genetically determined cancer. In: Genetic analysis of tumour suppression. Wiley, Chichester (Ciba Found Symp 142) p 3–19

Lee W-H, Bookstein R, Hong F, Young L-J, Shew J-Y, Lee EY-HP 1987 Human retinoblastoma susceptibility gene: cloning, identification and sequence. Science (Wash DC) 235:1394–1399

Lewis EB 1945 The relation of repeats to position effect in *Drosophila melanogaster*. Genetics 30:137–166

Mechler BM, McGinnis W, Gehring WJ 1985 Molecular cloning of *lethal(2)giant larvae*, a recessive oncogene of *Drosophila melanogaster*. EMBO (Eur Mol Biol Organ) J 4:1551–1557

Opper M, Schuler G, Mechler BM 1987 Hereditary Suppression of *lethal(2)giant larvae* malignant tumor development in *Drosophila* by gene transfer. Oncogene 1:91–96

Ruoslahti E, Pierschbacher MD 1986 Arg-Gly-Asp: a versatile cell recognition signal. Cell 44:517–518

Stanbridge EJ 1976 Suppression of malignancy in human cells. Nature (Lond) 260:17–20

## DISCUSSION

*Cavenee:* There was a report that the *l(2)gl* gene was homologous to the gene for the neural cell adhesion molecule N-CAM, is that true?

*Mechler:* I disagree with the conclusion drawn by Schmidt and his collaborators (Lützelschwab et al 1987).

*Marshall:* But there is an RGD motif in the *l(2)gl* gene indicating homology to the fibronectin class of adhesion proteins, isn't there?

*Mechler:* An RGD motif is present in the C-terminal portion of the p127 protein in a region which is not absolutely required for rescuing the *l(2)gl*-deficient animals. The homology with N-CAM is extremely weak and can be considered as statistically fortuitous. Furthermore, on a biological basis, if *l(2)gl* were related to N-CAM, one would expect that the absence of such an adhesion protein would lead to cell disaggregation at the end of embryogenesis; this does not happen. The development of the *l(2)gl*-deficient embryos proceeds normally. I would stress that the absence of *l(2)gl* has no direct effect on the development of the embryo. This is in contrast with most of the other genes expressed during early embryogenesis, like the segmentation and homeotic genes. The inactivation of these genes leads to an arrest of embryonic development.

*Nossal:* Lee Hood makes the point that in order to be a card-carrying member of the immunoglobulin supergene family, it is not so much a question of homology at the amino acid level, which in some cases may be only 20 or 25%, but the position of these Ig-like domains of 100 or so amino acids joined by disulphide bonds. By that criterion is *l(2)gl* a member of the immunoglobulin supergene family?

*Mechler:* By the position of the disulphide bridges, the *l(2)gl* gene does not correspond to a member of the immunoglobulin supergene family.

*Harris:* Alan Williams referees many papers that claim to have identified members of the immunoglobulin supergene family. He is always finding domains where people have missed them, if they are not used to looking for that particular concatenation of features that enables them to diagnose a part of the immunoglobulin gene. Figures for overall percentage homologies are usually uninformative.[1]

*Bodmer:* Have you looked under low stringency to see if there is any cross-hybridization of genes from other species, including mammals?

*Mechler:* Yes, we have isolated a large series of clones, sequenced many and found no real homology.

*Cavenee:* Has the dispensable part at the 3′ end of the gene been found to be lost in any of the naturally occurring *l(2)gl* mutants?

*Mechler:* No, we haven't found such a mutant.

*Cavenee:* How many naturally occurring alleles are there?

*Mechler:* More than 200 spontaneous or EMS-induced *l(2)gl* mutants have been isolated. We have examined 60 of them and found that all except two of these mutants were terminal deletions of the chromosome 2L. However, in all cases, the gene is lost or so disrupted that no defective *l(2)gl* proteins would be produced. We are now constructing transposons with mutations in the coding part of the *l(2)gl* gene so that we can obtain defective *l(2)gl* proteins. With such mutants, we will then be able to investigate which other proteins are interacting with the *l(2)gl* gene product.

*Wyke:* Earlier the paradox was brought up with the *Rb* gene—that it is a fairly universally expressed gene but implicated in only a narrow range of tumours. From your ring chromosome experiments you could perhaps explain this rather limited tumour range, but have you replaced one paradox with another? You now have the problem of a gene whose expression is very high in certain tissues at a later stage but its loss has no apparent consequence for the physiology of that tissue. What are your thoughts on that?

*Mechler:* My thought is that undoubtedly the *l(2)gl* proteins are interacting with other gene products and probably act in a cooperative fashion. Consequently, the absence of the *l(2)gl* proteins may produce different phenotypes, depending on the spatial and temporal expression of other genes. In *Drosophila*, genes that behave in a similar fashion are known. For example, *fushi tarazu* (*ftz*), a segmentation gene that controls cell identities during early pattern formation, has been shown to control cell fate during neurogenesis. Out of 250

---

[1]Williams has since examined the amino acid sequence of the *lethal 2* gene and not found any features to suggest that it is a member of the immunoglobulin supergene family.

neurons present in each hemisegment, the *ftz* gene is expressed in about 30 cells. In the absence of *ftz* expression, however, Corey Goodman's group (Doe et al 1988) have found that only one neuron displays a phenotypic change in an axon projection; the other cells look normal. A similar pattern is also found with *l(2)gl*. All embryonic cells express *l(2)gl*, but only a subset of cells shows a pattern of damage and an even smaller subset of cells becomes tumorous. Furthermore, the results of our mosaic experiment show that the establishment of tumorigenesis requires the complete absence of *l(2)gl* products. For me, these results indicate that *l(2)gl* interacts with other gene(s) or gene product(s). Our present goal is to identify these genes and gene products.

## References

Doe CQ, Hiromi Y, Gehring WJ, Goodman CS 1988 Expression and function of the segmentation gene *fushi tarazu* during *Drosophila* neurogenesis. Science (Wash DC) 239:170–175

Lützelschwab R, Klämbt C, Schmidt O 1987 A protein product of the Drosophila recessive tumor gene, l(2) giant gl, potentially has cell adhesion properties. EMBO (Eur Mol Biol Organ) J 6:1791–1797

# Suppression of genetic melanoma in the fish *Xiphophorus*

Manfred Schwab

*Institut für Experimentelle Pathologie, Deutsches Krebsforschungszentrum, D-6900 Heidelberg, FRG*

*Abstract.* Hybrids between certain species of the teleost *Xiphophorus* predictably develop melanomas. Classical Mendelian crossing experiments have allowed us to identify genetic loci involved in eliciting and in suppressing tumorigenesis. The overall picture is that melanoma formation results from functional elimination of a suppressor locus allowing abnormal expression of a melanoma locus. Melanoma formation in *Xiphophorus* behaves like a recessive trait. The fish melanomas consist of incompletely differentiated pigment cells and in many aspects resemble their murine and human counterparts. Cytogenetic studies of cells of the genetic melanomas have provided evidence for chromosomal abnormalities. In particular, cytogenetic manifestations of amplified DNA were detected in a cell line derived from a malignant melanoma. Amplified DNA was isolated and was found to be amplified in a particular type of genetic melanoma. Our results suggest that genetic changes in addition to elimination of suppressor genes contribute to the malignant phenotype of melanoma in *Xiphophorus*. It is possible that a similar situation occurs in cancers of higher vertebrates, including humans.

*1989 Genetic analysis of tumour suppression. Wiley, Chichester (Ciba Foundation Symposium 142) p 181–198*

Recent years have brought considerable advances in our understanding of the contributions of cellular genes to carcinogenesis. Studies of sporadic cancers have identified a wide range of proto-oncogenes in which mutation or dysregulated expression resulting from relocation or amplification appear to be cardinal factors in carcinogenesis. Analyses of cancers with a genetic background have also pointed to deletion or functional loss of genetic material in some cancers. These observations have led to the concept that the genome contains a group of genes responsible for maintaining the normal phenotype of the cell and suppressing carcinogenesis. The operational term tumour suppressor genes was given to genes having this function. The relative importance to carcinogenesis of each of the genetic changes involving proto-oncogenes or tumour suppressor genes is not clear for most cancers. There is general agreement, though, that development of cancer is a multistep process with each of the steps characterized by a cell population differing qualitatively from the others, presumably as the result of one or several genetic changes. It is also possible that even those genetic changes described as 'dominant' remain phenotypically silent in specific developmental situations or within a given tissue architecture and contribute to malignant change only in concert with a particular genetic alteration involving different information.

Genetic melanoma in offspring of interspecific crosses in the fish *Xiphophorus* is prototypic for the concept of cancer being a genetic disease at the cellular level. Here the autosomal suppressor gene elimination which results in abnormal expression of an X-chromosomal oncogene is a key event eliciting carcinogenesis. It can reproducibly be shown in a vertebrate, above the level of somatic cell hybrids, that carcinogenesis is normally suppressed. Significantly, the genetic cancers are not restricted to laboratory mutants but result from crosses of wild-type fish as well. Based on the executive role that cellular genes have for carcinogenesis in *Xiphophorus* we believe that higher vertebrate genes function in analogous processes. In this paper our present view about oncogenes and tumour suppressor genes in *Xiphophorus* is discussed.

## Genetic melanoma

Melanoma is the tumour that arises spontaneously in *Xiphophorus* in a Mendelian fashion and most of the information on genetic suppression of malignancy has been generated by studying this particular type of tumour. In general we can recognize two classes of genes involved in tumorigenesis: 1) oncogenes, that make a positive contribution to the malignant phenotype; and 2) tumour suppressor genes, that make a positive contribution to the normal phenotype. Animals wild-type for both oncogenes and tumour suppressor genes have a normal phenotype. Tumorigenesis can be elicited when the genetic balance between the oncogenes and the tumour suppressor genes is disturbed in favour of the oncogene, or when the oncogene is altered in such a way that it eludes the cellular control mechanisms of tumour suppressor genes (for review, see Schwab 1987b).

There is a series of genes in *Xiphophorus* that formally behave like oncogenes. These include the macromelanophore loci, like *'spotted dorsal' (Sd)* and *'spotted'* (*Sp*) (Gordon 1959, Kallmann 1975, Anders 1967). There seem to be at least two genetic loci that behave like tumour suppressor genes, *'Differentiation'* (*Diff*), and *'golden'* (*g*). There are several reviews available on the *Xiphophorus* tumour model (see Gordon 1927, 1959, Anders 1967, Kallmann 1975, Ozato & Wakamatsu 1983, Anders et al 1984, Schwab 1986a, 1987b, Vielkind & Vielkind 1982). As the interpretations of the genetic results differ somewhat among the authors, our present view is outlined here.

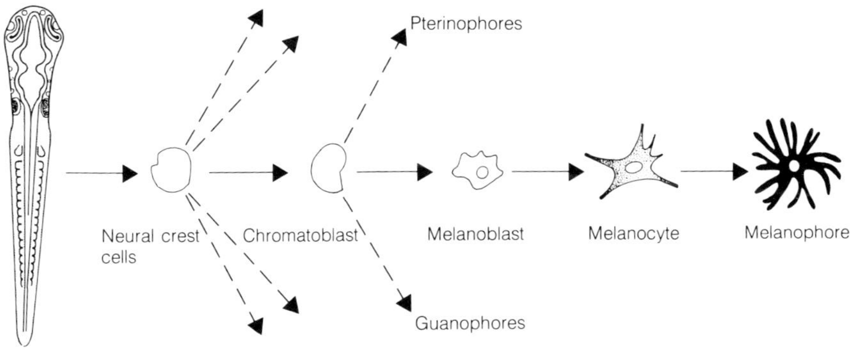


FIG. 1. Schematic illustration of melanophore differentiation.

## Oncogenes as a prerequisite for melanoma development

Individuals of the different species of *Xiphophorus* display characteristic spot patterns composed of cells containing the pigments melanin (melanophores), pteridin (pterinophores) and guanin (guanophores). Among the melanophores at least two major types have been distinguished: 1) micromelanophores with an average size of about 30 μm, and 2) macromelanophores, giant cells up to 2 mm in diameter (pictured in Schwab 1987b).

Differentiation of melanophores starts after migration of cells from the neural crest of the early embryo. The chromoblast appears to be the common precursor of all pigment cells of the fish, including melanophores, pterinophores and guanophores (Bagnara et al 1979) (Fig. 1). Melanophores differentiate via the stages of melanoblast and melanocyte. It is not clear if macromelanophores and micromelanophores develop from common precursor cells or represent different lines of development.

Micromelanophores are present on all wild-type animals and are distributed more or less evenly in the dermis, although in fish of some species they congregate to form small spots. Macromelanophores are displayed only by certain species, and they are generally arranged in groups to form conspicuous spots. *Xiphophorus maculatus* (a member of the 'platy' group), which can be viewed as the prototype species carrying macromelanophores, can display two different types of macromelanophore pattern: one in the dorsal fin (Spotted dorsal; Sd), the other on the side of the body (Spotted; Sp) (Fig. 2a). Development of the two patterns is directed by the corresponding X-linked loci, *Sd* and *Sp*. No recombination has so far been observed between *Sd* and *Sp*, suggesting that the loci are very closely linked on the X chromosome. *Sd* and *Sp* may even be alleles at the same locus but as yet there is no formal proof of this. In fish of all genotypes the oncogenes are expressed

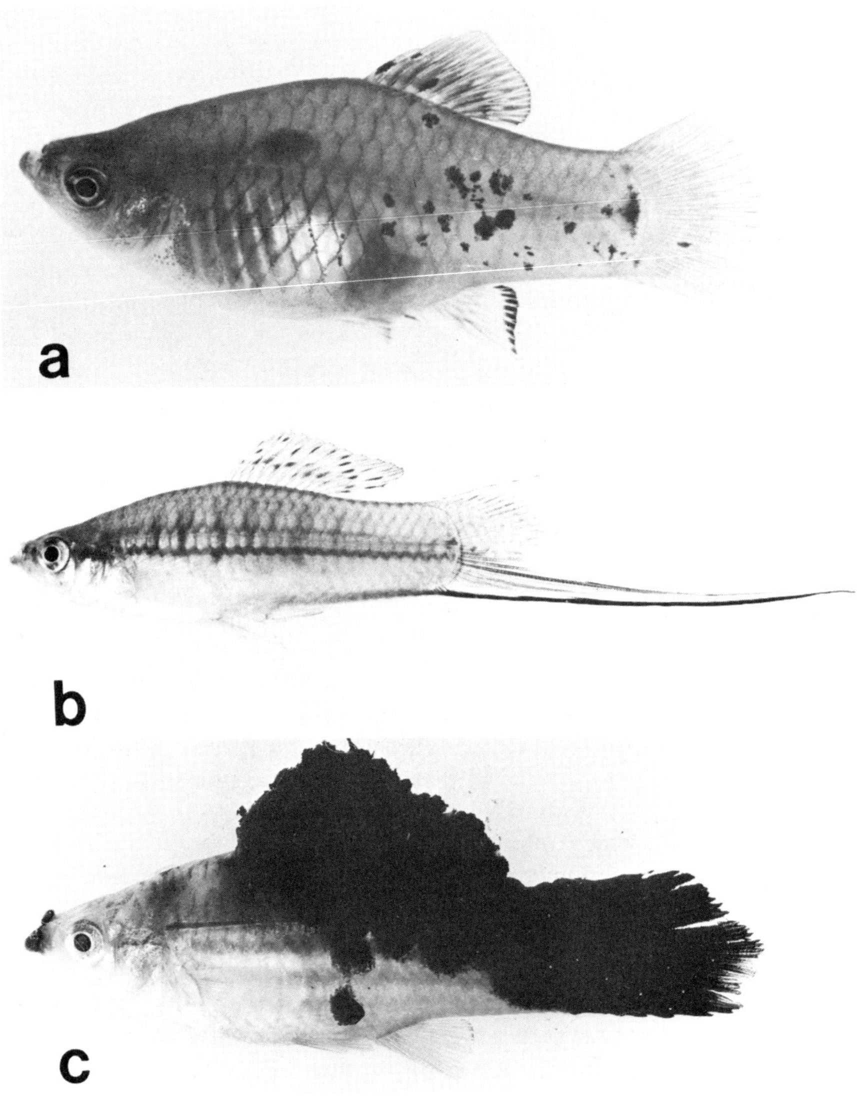


FIG. 2. Genetic melanoma. Crossing of *X. maculatus* (a) with *X. helleri* (b) and backcrossing of the interspecies F1 yields malignant melanoma in backcross segregants, *BC*(c). See also Fig. 3.

co-dominantly. Individuals homozygous for either of the two loci display only the pattern corresponding to that locus.

Crossing of *SdSp* platy with a strain of *X. helleri* ('swordtail') (Fig. 2b) that does not display any spot pattern yields F1 animals that show either the Sd or the Sp pattern, in accordance with the segregation of the two loci. The area covered by melanophores is considerably larger than that in the parental platy, but this hyperpigmentation should not be regarded as a melanoma. When fish of the F1 generation, which are fully fertile, are backcrossed to the swordtail, the expected 1:1 segregation with respect to the macromelanophore locus *Sd* of *Sp* is seen. Fish lacking *Sd* or *Sp* phenotypically resemble the swordtail. Fish carrying *Sd* or *Sp* develop more or less extensive hyperpigmentation (melanosis), or display large melanomas (Fig. 2c). Strains that are homozygous for the autosomal locus '*albino*' have unpigmented melanomas. In general, when genes normally directing the development of macromelanophore patterns are introduced into a specific heterologous genetic environment a condition arises in which pigment cell differentiation goes awry and melanomas develop instead.

Genetically linked to *Sd* but apparently not to *Sp* is a locus '*Dorsal red*' (*Dr*) directing development of pterinophores in the dorsal fin. A strain has been bred that develops the Dr phenotype, but lacks macromelanophores: crossings involving these fish fail to yield fish with melanomas. This finding clearly shows that the functional oncogene must be present for melanoma to develop.

## Tumour suppressor genes

At least two genetic loci have been identified that suppress formation of melanomas: *Differentiation* (*Diff*) and *golden* (*g*).

*Differentiation*. Genetic studies involving specific mutants revealed that approximately 50% of the backcross segregants carrying the macromelanophore locus had a benign hyperpigmentation, while the remaining 50% had a fully fledged melanoma. The same phenomenon was also seen when wild-type fish were used in crossing experiments (Fig. 3). This 1:1 segregation indicated that an autosomal locus transmitted from the platyfish was involved in determining the degree of malignancy. Because a high degree of pigment cell differentiation was the primary correlate to this locus, the putative gene was tentatively designated *Diff*. The idea that activity of *Diff* was involved in suppression of the malignant phenotype gained support when surveys of isozymes revealed a polymorphic esterase, Esterase-1, encoded by a gene linked to the putative *Diff* locus (Ahuja et al 1980, Schwab & Scholl 1981). Interestingly, in humans esterase D is genetically linked to the retinoblastoma gene, which like *Diff* can be regarded as a tumour suppressor gene (for review, see Hansen & Cavenee 1987), but this provocative parallel is presumably coincidental. Esterase-1 and the benign pigment cell phenotype co-segregate in crossings involving *X. maculatus* and *X. helleri* (Fig. 3). These observations suggest that the macromelanophore locus is normally suppressed

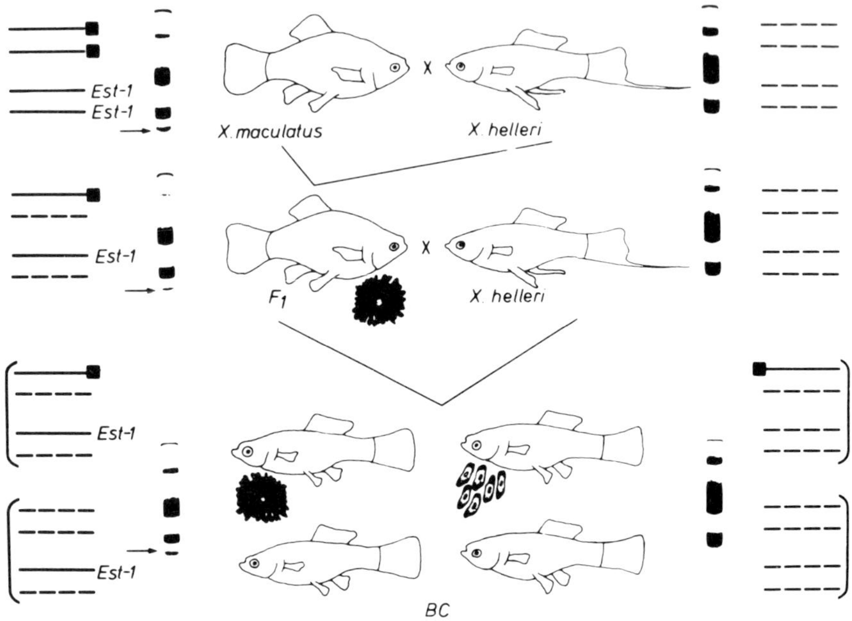


FIG. 3. Tumour suppressor gene *Diff*. *Diff* is linked to *Est-1* on the same autosome. Introduction of the X-chromosomal melanoma gene transmitted from *X. maculatus* (indicated by a line with a solid square) into the genetic background of *X. helleri* (broken lines) results in segregation of *Diff* and the melanoma gene (only two of the 24 chromosome pairs of *Xiphophorus* are considered here). *Est-1* is correlated with the benign phenotype, segregants with malignant melanoma usually lack *Est-1*. *BC*, backcross segregants.

in *X. maculatus* by a *trans*-acting locus, designated *Diff*. Segregation of the X-linked macromelanophore and the autosomal *Diff* loci during backcrossing results in fish carrying malignant melanoma. Activation of the macromelanophore loci *Sd* and *Sp* to direct melanoma formation appears to be the result of elimination of a tumour suppressor gene.

*Golden*. The autosomal recessive locus *golden* almost completely abolishes the development of melanophores. Backcross segregants homozygous for *golden* and carrying the macromelanophore locus *Sd* or *Sp* do not show hyperpigmentation or melanoma (Schwab 1987b). Little is known about the cellular mechanisms governing lack of melanophores but homozygosity for *golden* results in a block in the differentiation of cells of the melanophore lineage at a stage before melanin synthesis has started. Thus it has been suggested that *golden* may indicate the presence of a tumour suppressor gene, since its elimination results in development of melanoma. The suppressor genes behave in a dosage–dependent manner: only double dosage will completely rescue the normal phenotype.

**Genetic suppression of susceptibility to cancer**

The Y chromosome of *X. maculatus* carries a macromelanophore locus *Striped* (*Sr*) that does not direct spontaneous formation in interspecies hybrids. The X-linked *Lineatus* (*Li*) and the Y-linked *Punctatus* loci of *X. variatus* (*Pu*) behave in a similar way (for review, see Anders et al 1984). Two questions arose; whether these loci can be activated to direct melanoma formation, and whether additional oncogenes in various hybrid combinations can be activated to direct development of other types of tumours.

*Susceptibility to melanoma*

By X-irradiation it was possible to obtain mutant germline versions of *Sr* and *Li* that, like *Sd* or *Sp*, are able to direct melanoma development in backcrosses involving *X. helleri* as the recurrent parent. Two such mutants have been particularly useful. One represents a translocation of the X-linked *Li* in the vicinity of the *Dr* locus (*DrLi*), the other a mutation of *Sr* in its usual locus (Anders et al 1984). It has been suggested that the activation of the oncogenic potential in these mutants results from mutational impairment of the regulatory control that is thought to be linked to the *Li* or *Sr* locus. However, this interpretation should be viewed with caution, since as yet there are no data supporting the hypothesis. Studies of other systems have shown that the oncogenic potential of cellular genes can also be activated by other mechanisms. In the present context, two mechanisms are particularly relevant: 1) point mutation within the coding region of a gene, as exemplified by the mutations in specific codons of the cellular Ha-*ras* gene; and 2) generation of a fusion protein, for instance in human chronic myelogenous leukaemia, where a translocation replaces part of the c-*abl* gene with coding sequences from the breakpoint cluster region, and a fusion protein is produced from the chimaeric gene (for review, see Bishop 1987).

Identification of the molecular mechanisms activating the oncogenic potential of the melanoma oncogenes that are refractory to activation by crossings represents an interesting challenge for future studies.

*Susceptibility to non-melanotic tumours*

Despite large scale interspecies breeding programmes in several laboratories, spontaneous non-melanotic tumours have been found extremely rarely. One could argue that the loci directing macromelanophore development are not essential and their regulatory control is evolutionarily less conserved than that of the genes involved in development of more vital cell systems. Initial studies involving X-irradiation to induce non-melanotic tumours in animals of wild-type stocks also failed. This observation can be taken as another indicator for tight control of normal cellular regulation in the fish. More recent studies have indicated, however, that differentiation of cells derived from various embryonal lineages is unstable in backcross hybrids (for review, see Schwab 1986a). Evidence comes from experiments in which non-hybrids, F1 and backcross hybrids were exposed to the mutagens/carcinogens N-methyl-N-nitrosourea or X-rays. While non-hybrids proved to be not susceptible to the development of tumours, F1 hybrids developed melanomas at a low rate of incidence. In contrast, backcross hybrids displayed a large spectrum of various types of tumours, including fibrosarcomas, melanomas, carcinomas of various organs and a retroocular tumour identified as a neuroblastoma. The usual explanation for the high susceptibility of backcross hybrids compared with non-hybrids is based mainly on the assumption that genes involved in growth control have evolved separately in different species of *Xiphophorus* (Schwab 1986a, 1987b). Crossing and backcrossing might result in a condition in which genetic mechanisms involved in growth control are impaired. Additional damage to growth regulation by carcinogenic/mutagenic agents might then trigger tumour development. The nature of the genetic factors responsible for high susceptibility in backcross hybrids is not known.

## Tumour suppression by cellular differentiation

Tumour suppression in interspecies hybrids of *Xiphophorus*, as well as in somatic hybrids involving human and murine cells, is defined genetically. A plethora of mechanisms able to suppress the malignant phenotype can be envisioned and it is rather difficult to develop a convincing understanding of how any of these might be related to cancer. Theoretically, suppression of malignancy can occur at any of the steps in gene expression, from transcription to post-translational interactions, or at the level of intercellular communication. At this point no biochemical function of a tumour suppressor gene has been identified.

In *Xiphophorus* there is some evidence for pleiotropic effects of the suppressor gene *Diff*. The major feature associated with *Diff* is cellular differentiation. Homozygous presence of *Diff* results in terminal differentiation of most of the macromelanophores, and in suppression of malignancy; lack of *Diff* prevents most of the macromelanophore precursor cells, i.e. melanoblasts and melanocytes, becoming terminally differentiated and melanoma develops (Fig. 4).

Detailed cytological and biochemical studies have yielded a reasonable definition of melanocytes and melanophores, in particular those developing under the direction of the *Sd* locus. The melanocyte has been defined as a

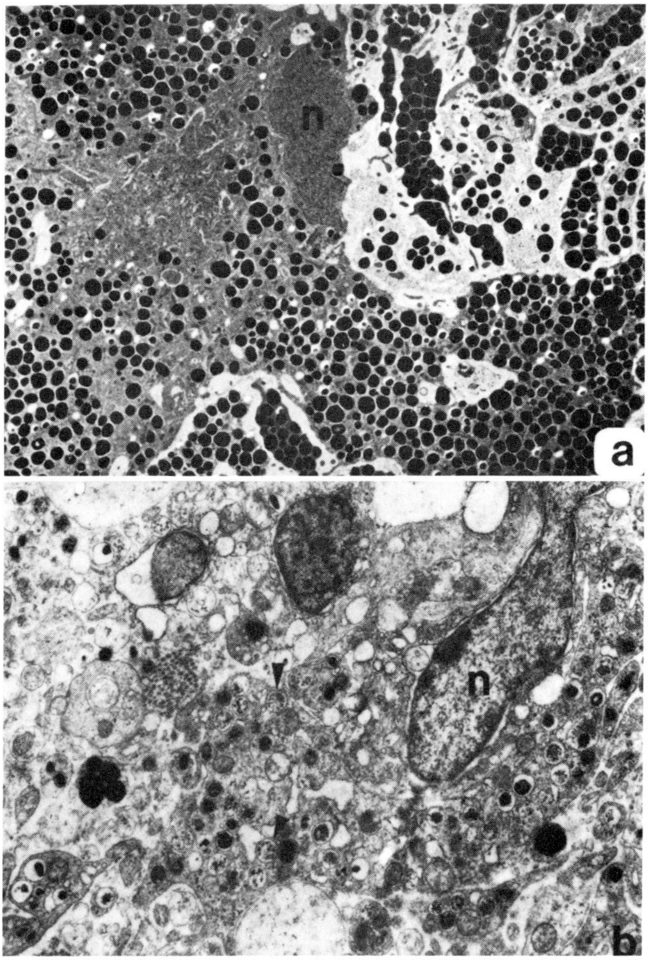


FIG. 4. Ultrastructure of a differentiated melanophore (a) versus an undifferentiated melanocyte (b) in backcross segregants (*Diff*$^+$*Diff*$^-$ and *Diff*$^-$*Diff*$^-$, respectively). Differentiated melanophores have stopped dividing; undifferentiated cells are actively dividing. n, nucleus. Arrow indicates pre-melanosome.

melanin-producing and melanin-containing cell. It is regarded as the lineal descendant of the unpigmented melanoblast, which originates from the neural crest of the developing fish embryo. The term 'melanocyte' includes both younger and older stages, the older stages containing more melanin than the younger. Melanocytes range from spindle-shaped to various dendritic forms. All melanocytes have the characteristics of typical proliferating pigment cells and have high rates of DNA synthesis. They usually possess well-developed endoplasmic reticulum and Golgi complexes, and all stages of developing melanosomes (the intracellular melanin-containing granules) can be found. In agreement with the high rate of melanin synthesis, tyrosinase activity is high (Vielkind 1976).

The melanophore has been defined as a pigment effector cell. It has the characteristics of a terminally differentiated, non-proliferating cell and shows virtually no DNA synthesis. Melanophores are highly dendritic. They contain mature melanosomes filled with melanin and lack other cytoplasmic structures, such as endoplasmic reticulum or Golgi complexes. Lack of tyrosinase activity indicates that melanin synthesis has stopped.

As a general rule, there is an inverse correlation between the malignancy of the melanoma and the degree of pigment cell differentiation. Melanomas consisting of young melanocytes form large masses and are highly invasive, while melanomas consisting predominantly of highly differentiated cells are considerably smaller and tend to be confined to their site of origin. The most rapidly growing melanomas develop in individuals that are homozygous for the autosomal locus *albino*. These amelanotic melanomas consist of very immature pigment cells that lack melanosomes altogether. Detailed histological examinations have shown that the malignant melanomas of *Xiphophorus* resemble in many aspects the melanomas found in the mouse or in humans.

*Diff* of *Xiphophorus* behaves very similarly to the *Rb* gene, the best characterized tumour suppressor gene in humans. Homozygous deletion or inactivation of *Rb* seems to be a prerequisite for retinoblastoma (see Knudson, this volume). The timing of development of retinoblastoma in children and the restricted tissues in which the homozygous functional loss of the *Rb* gene results in tumorigenesis strongly suggest that *Rb*, like *Diff*, may be a differentiation-controlling gene. Loss of *Rb*, like that of *Diff*, may prevent terminal differentiation of already committed cells and thereby permit further unrestrained proliferation.

## Genetic reversion of the malignant phenotype

The genetic studies of melanoma development in offspring of normal fish suggest that melanomas result from abnormal gene regulation in pigment cells of interspecies hybrids. There is no indication that interspecies hybridization results in general chromosomal instability in the hybrids or in genetic alterations. The major genetic change involved in tumorigenesis appears to be segregation of the melanoma oncogene from the *Diff* suppressor gene. It is obvious therefore, to test if reintroduction of the *Diff* gene into the genetic environment of the melanoma locus, for instance *Sd*, will result in abrogation of malignancy. This can be tested by crossing an $Sd^+Sd^-Diff^-Diff^-$ backcross hybrid carrying malignant melanoma (Fig. 5a) with an $Sd^+Sd^-Diff^+Diff^+$ *X. maculatus* (Fig. 5b). Because the sex determining factor on the *Sd* chromosome

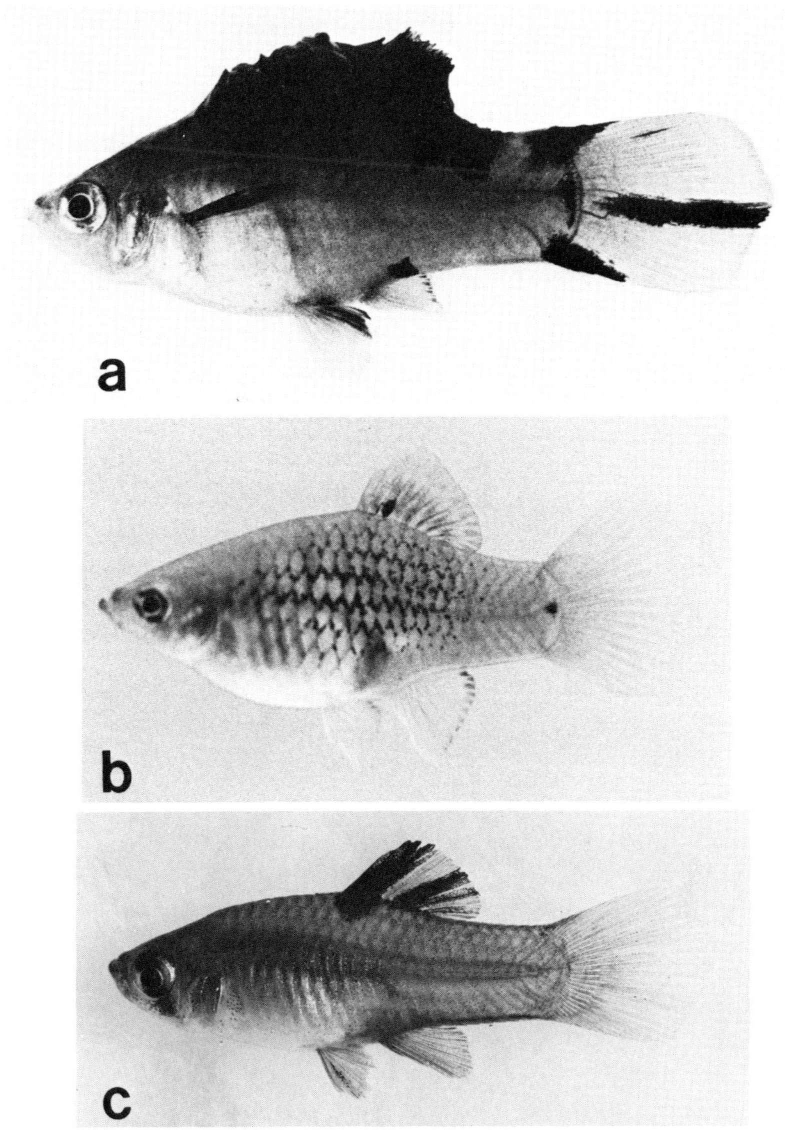

FIG. 5. Reversion of the malignant phenotype. Crossing of *BC* segregant carrying malignant melanoma (a) with normal *X. maculatus* (b) yields phenotypically uniform offspring without malignant melanoma (c; see Fig. 6 for genetic interpretation).

usually dictates female development, most *BC* hybrids are female. Consequently, the crossing partner *X. maculatus* must be male, have only one copy of the *Sd* chromosome (the Y chromosome carries a different oncogene)

| | | *Xiphophorus maculatus* $Sd^+\ Sd^-;\ Diff^+\ Diff^+$ | |
|---|---|---|---|
| Backcross hybrid with malignant melanoma $Sd^+\ Sd^-;$ $Diff^-\ Diff^-$ female | Gametes | $Sd^+\ Diff^+$ | $Sd^-\ Diff^+$ |
| | | A | C |
| | $Sd^+\ Diff^-$ | $Sd^+\ Diff^-$<br>$Sd^+\ Diff^+$ | $Sd^-\ Diff^+$ |
| | | B | D |
| | $Sd^-\ Diff^-$ | $Sd^-\ Diff^-$<br>$Sd^+\ Diff^+$ | $Sd^-\ Diff^+$ |

FIG. 6. Reversion of the malignant phenotype.

and be homozygous for the autosomal *Diff* (Fig. 6). In practice, such crosses can be performed only before melanoma formation becomes too severe. The offspring of such crosses carrying the *Sd* locus (genotypes A, B, C in Fig. 6) show hyperpigmentation, resemble the F1 between *X. maculatus* and *X. helleri* and do not show melanoma. It appears that reintroduction of genes from *X. maculatus*, presumably the differentiation-promoting gene *Diff*, results in suppression of the malignant phenotype. As yet no revertants showing the malignant phenotype have emerged from such crosses.

### DNA amplification and suppressor elimination — genetic changes in malignancy

Gene amplification refers to the selective increase of the copy number of cellular genes. The molecular basis for gene amplification is poorly understood, although some ideas have been proposed (Schimke 1984, Stark 1984). Briefly, there seems to exist in the cell a certain degree of spontaneous illegitimate DNA replication, resulting in replication of portions of the replicon more than once during a single cell cycle. It is germane to the process of tumour progression that this rate can be increased severalfold by various experimental drugs, including tumour promoters. Normally the extra replicated DNA is lost, but if there is a selective advantage for its retention, it is stabilized.

Cytogenetic studies of human and animal tumour cells have repeatedly provided evidence for mysterious chromosomal manifestations of amplified cellular DNA: 'double minutes', 'C-bandless chromosomes', and 'homogeneously staining chromosomal regions', but their origin, genetic information and possible significance remained enigmatic for a long time (for review, see Alitalo & Schwab 1986). During the past years it has emerged that these chromosomal features are frequently the sites of amplified cellular oncogenes in cancer cells (reviews, Alitalo & Schwab 1986, Schwab 1985, 1986b, 1987a).

Because DNA amplification is a common mechanism for oncogenic activation of cellular genes it was obvious to test whether cellular DNA is amplified in *Xiphophorus* tumour cells. Preliminary cytogenetic studies by Wakamatsu (1981) had indicated that *in vitro* cultivated cells (PSM-1) of an albino *Sp* tumour of a backcross hybrid involving *X. maculatus* and *X. helleri* contain 'double minutes'. These observations are difficult to document due to the small size of chromosomes present in *Xiphophorus* cells. Our own cytogenetic studies indicated the presence of homogeneously staining chromosomal regions (Silke Brüderlein, unpublished data). Molecular analyses aimed at identifying whether a known cellular oncogene had been amplified did not produce positive results (unpublished data). By subtractive cloning we have identified and cloned amplified DNA in PSM-1 cells (Schwab et al 1989). Our preliminary experiments suggest that this DNA is a portion of a gene expressed in a developmentally regulated pattern and that this gene is highly conserved from fish to humans.

Formal genetics has indicated that cancer development in *Xiphophorus* starts when oncogenes are abnormally expressed due to elimination of tumour suppressor genes. Physical or functional loss of the oncogenes is associated with loss of ability to develop melanoma even when the suppressor gene *Diff* is eliminated. *Diff* is not necessarily directly involved in regulation of oncogene expression: it probably has a more general function related to differentiation of pigment cells. Loss of *Diff* function abrogates this differentiation. The immature stage of differentiation could be related to constitutive oncogene expression that results in abnormal expansion of the population of immature pigment cells. It will be difficult, though, to sort out the exact contributions of oncogenes and suppressor genes to malignancy, even in the genetically well defined model of *Xiphophorus*. Cellular functions of tumour suppressor genes may turn out to be as complex and pleiotropic as those of oncogenes.

The gene *Diff* appears to suppress malignancy by directing terminal differentiation of pigment cells: this may also be one of the properties of human tumour suppressor loci, in particular the *Rb* gene. Although it is difficult at this point to envisage which molecular or biochemical function of tumour suppressor genes we might be able to identify, research on tumour suppression will at least illuminate basic mechanisms of cell differentiation and multiplication. In *Xiphophorus* it has become clear that a particular suppressor locus is involved in the differentiation of a specific cell lineage and elimination from the cell results in cancer of only this particular cell lineage. From the specificity with which DNA is deleted in particular human cancers it appears reasonable to assume that corresponding mechanisms operate in humans. Chromosomal deletions presumably signify genetic loci involved in the differentiation programme of specific cell lineages. It is not clear, though, if elimination of tumour suppressor genes alone is sufficient to elicit the fully malignant phenotype. Cytogenetic studies have shown various non-random chromosomal abnormalities in those human tumours in which elimination of a tumour suppressor gene seems to be a critical step in tumorigenesis. In *Xiphophorus*, it is obvious from our molecular studies that additional genetic events, such as DNA amplification, can contribute to the malignant phenotype of the cell.

At present, we can only speculate about the role that amplification might play in melanoma development. The development of a malignant melanoma in *Xiphophorus* appears to be a dynamic process. It is a common observation that a particular melanoma phenotype is stable for a long time, after which a new growth phase starts, leading to expansion of the melanoma. These observations indicate that new pigment cell phenotypes arise that possess more malignant characteristics. From studies in humans and animals it is known that this progression phase is associated with major genetic changes in the cellular genome and these changes are indicated by chromosomal abnormalities. It is possible that amplification of cellular DNA, as detected in melanoma cells of *Xiphophorus*, could be related to malignant progression. Normally a high percentage of melanocytes, even when transformed, appears to undergo terminal differentiation to melanophores. These, after becoming senescent, may die off. They will be replaced by newly differentiating cells. In animals of the *Diff*/− genotype, this dynamic process favours the melanophores, resulting in hyperpigmentation. In contrast, in animals lacking *Diff* it favours melanocytes and malignant melanomas develop instead. It is our hypothesis that DNA amplification could result in the emergence of new populations of melanocytes that have a low tendency to differentiate to melanophores (Fig. 7). Amplification therefore could also be responsible for development of malignant melanomas in animals of the *Diff*/− genotype. In earlier studies some backcross segregants carrying malignant melanoma were found to be positive for *Esterase-1*, which is linked to *Diff* (Ahuja et al 1980,

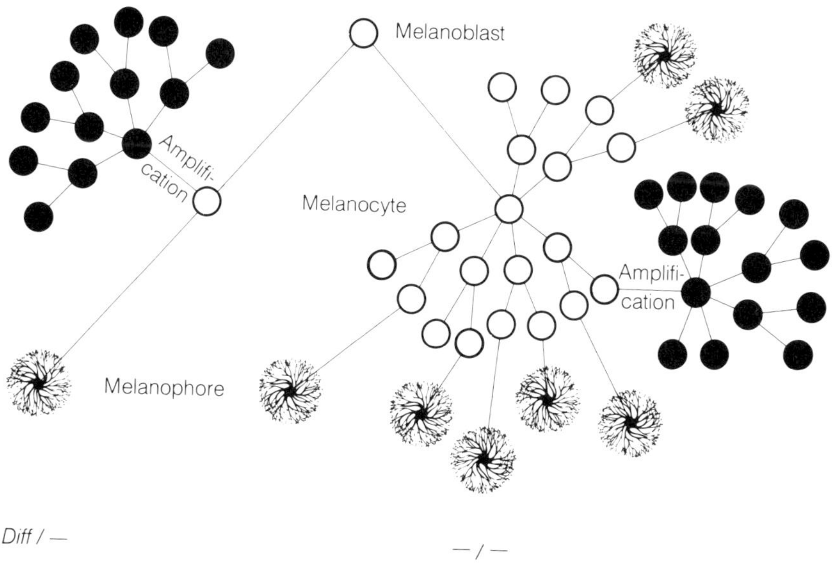


FIG. 7. Possible contribution of amplification to tumour progression in melanomas of *Xiphophorus*. According to this hypothesis, amplification can contribute to progression both in *Diff*$^+$*Diff*$^-$ and in *Diff*$^-$*Diff*$^-$ segregants. Normally, in *Diff*$^+$*Diff*$^-$ segregants most transformed melanocytes will differentiate terminally to melanophores. This process results in benign neoplasia. Amplification in melanocytes could interfere with differentiation and lead to malignant melanoma. In *Diff*$^-$*Diff*$^-$ segregants normally only a few melanocytes will differentiate to melanophores; most melanocytes will form a malignant melanoma. The formation of melanoma can be accelerated after amplification, resulting in a population of cells unable to differentiate to melanophores.

Schwab & Scholl 1981). These results were explained by loose linkage between *Est-1* and *Diff*, resulting in individuals carrying *Est-1* but lacking, due to recombination, *Diff*. It is possible that the individuals were carrying both *Est-1* and *Diff*, and that amplification of DNA had resulted in *Diff*-carrying animals in malignant melanoma cells. In conclusion, amplification is thought to contribute to increased growth of the already neoplastic phenotype, i.e. it is a third genetic contributor in addition to the melanoma oncogene and the *Diff* suppressor gene.

## Summary

Genetic tumours of *Xiphophorus* are among the classical experimental models which underline the concept that cancers develop as a result of abnormal gene expression. Formal genetics has indicated that cancer development in

*Xiphophorus* starts when oncogenes are abnormally expressed as the result of elimination of tumour suppressor genes, although there does not appear to be a direct regulatory interaction between the oncogene and the suppressor gene. It is obvious from our molecular studies that additional genetic events can contribute to the malignant phenotype. Among these, amplification of cellular DNA may play a role in malignant progression of melanomas. At present, the exact contribution that amplification makes to the genetic melanoma is unclear. Judging from the role of amplification that has emerged from studies of human and murine tumours, amplification in melanomas of *Xiphophorus*, in addition to suppressor elimination, is likely to be significant.


### *Acknowledgements*

I thank Ingrid Ulbrich for typing the manuscript and Carmen Kippenhahn for art work.


## References


Ahuja MR, Schwab M, Anders F 1980 Linkage between regulatory locus for melanoma cell differentiation and an esterase locus in *Xiphophorus*. J Hered 71:403–407

Alitalo K, Schwab M 1986 Oncogene amplification in tumor cells. Adv Cancer Res 47:235–279

Anders F 1967 Tumor formation in platyfish — swordtail hybrids as a problem of gene regulation. Experientia (Basel) 23:1–10

Anders F, Schartl M, Barnekow A, Anders A 1984 *Xiphophorus* as an *in vivo* model for studies on normal and defective control of oncogenes. Adv Cancer Res 42:191–275

Bagnara JT, Matsumoto J, Ferris W, Frost S, Turner WA, Tchen TT, Taylor JD 1979 Common origin of pigment cells. Science (Wash DC) 203:410–415

Bishop JM 1987 The molecular genetics of cancer. Science (Wash DC) 235:305–311

Gordon M 1927 The genetics of a viviparous top-minnow *Platypoecilius* the inheritance of two kinds of melanophores. Genetics 12:253–283

Gordon M 1959 The melanoma cell as an incompletely differentiated pigment cell. In: Gordon M (ed) Pigment Cell Biology. Academic Press, New York, p 215–239

Hansen MF, Cavenee WK 1987 Genetics of cancer predisposition. Cancer Res 47:5518–5527

Kallmann KD 1975 The platyfish *Xiphophorus maculatus*. In: King RC (ed) Handbook of Genetics. Plenum Publishing Corporation, New York vol 4:81–132

Knudson AG 1989 Epidemiology of genetically determined cancer. In: Genetic analysis of tumour suppression. Wiley, Chichester (Ciba Found Symp 142) p 3–19

Ozato K, Wakamatsu Y 1983 Multi-step genetic regulation of oncogene expression in fish hereditary melanoma. Differentiation 24:181–190

Schimke RT 1984 Gene amplification in cultured animal cells. Cell 37:705–713

Schwab M 1985 Amplification of N-*myc* in human neuroblastomas. Trends Genet 1:271–275

Schwab M 1986a Malignant metamorphosis: developmental genes as culprits for oncogenesis in *Xiphophorus*. Adv Cancer Res 47:63–97

Schwab M 1986b Amplification of proto-oncogenes and tumor progression. In: Kahn P, Graf T (eds) Oncogenes and growth control. Springer Verlag, Heidelberg, p 332–339

Schwab M 1987a Amplification of cellular oncogenes during evolution of tumor cells. In: zur Hausen H, Schlehofer J (eds) The role of DNA-amplification in carcinogenesis. Lippincott, Philadelphia, p 178–185

Schwab M 1987b Oncogenes and tumor suppressor genes in *Xiphophorus*. Trends Genet 3:15–18

Schwab M, Scholl E 1981 Neoplastic pigment cells induced by N-methyl-N-nitrosourea (MNU) in *Xiphophorus* and genetic control of their terminal differentiation. Differentiation 19:77–83

Schwab M, Oehlmann R, Brüderlein S, Wakamatsu Y 1989 Amplified DNA in cells of genetic melanoma of *Xiphophorus*. Oncogene, in press

Stark GR 1984 Gene amplification. Annu Rev Biochem 53:447–491

Vielkind U 1976 Genetic control of cell differentiation in platyfish-swordtail melanomas. J Exp Zool 196:197–204

Vielkind J, Vielkind U 1982 Melanoma formation in fish of the genus *Xiphophorus*: a genetically-based disorder in the determination and differentiation of a specific pigment cell. Can J Genet Cytol 24:133–149

Wakamatsu Y 1981 Establishment of a cell line from the platyfish-swordtail hybrid melanoma. Cancer Res 41:679–680

## DISCUSSION

*Varmus:* In how many tumours are the same genes amplified?

*Schwab:* We have looked at only one gene. The degree of amplification is correlated with the different growth behaviours of the clones, for instance, growth in soft agar, but these genes don't seem to be amplified to the same extent in the various clones. We are concentrating on one gene, for which there seems to be selection in those cases of better growth. As far as the tumours are concerned, we have looked at five or six tumours and have found amplification of this particular gene in two or three—but that doesn't mean much. Amplification could be a feature of only a few cells, or it could be confined to certain areas of the tumour. So if we look at total DNA, we may fail to detect such amplification.

*Benedict:* This reminds me of the neuroblastoma story where N-*myc* amplification seems to be involved in the aggressiveness of the tumour more than in the aetiology of the tumour. I am trying to think of a known case in which the amplification of a specific gene has been correlated with the aetiology of the tumour rather than with some other phenomenon. It is a big step from finding amplification of a particular gene in a human cell to showing that this change is important in the tumorigenicity of that cell.

*Schwab:* I have several ideas about why this particular amplification may not be a prominent mechanism in the early stages. All the cases in which people have found amplification tend to correlate the amplification with more aggressive behaviour, i.e. more vigorous cell growth. As far as tumorigenicity is concerned, the correlation is with the progression phase, which is a late phase. I think it is fair to assume in melanoma in the fish that the early stages of tumorigenesis result from the elimination of information, rather than from a genetic or mutational change.

*Harris:* You can get gene amplification of one kind or another whenever you have selection, whether it is conscious selection or not. For example, you can get substantial amplification of the c-*ras* genes in diploid fibroblasts, if you grow them long enough. The amplification is most marked when the cells become senescent. I think these gene amplifications are probably always the result of selective pressures that occur during the growth of these cells *in vitro* or *in vivo*. I don't know of a single plausible case for a gene amplification being the initiating event in determining malignancy.

*van der Eb:* I assume that you can't grow these tumour cells in nude mice.

*Schwab:* We are thinking about it. They seem to grow but the problem is the temperature.

# The biology of tumour suppression

Henry Harris

*Sir William Dunn School of Pathology, University of Oxford, South Parks Road, Oxford OX1 3RE, UK*

*Abstract.* Evidence is presented for the view that suppression of malignancy in hybrids between malignant and normal cells is achieved in one of two ways. The normal cell either imposes its own pattern of terminal differentiation on the hybrid or it complements a genetic block in the differentiation programme of the malignant cell. The progressive multiplication of malignant cells in the animal is seen as a secondary consequence of blocked differentiation. Karyological examination of hybrids in which malignancy is suppressed is entirely consistent with the view that programmes of differentiation may be governed in a pleiotropic fashion by single genes.

*1989 Genetic analysis of tumour suppression. Wiley, Chichester (Ciba Foundation Symposium 142) p 199–213*

Although it has taken almost twenty years to achieve an amicable consensus of opinion, it is now agreed that when malignant cells are fused with normal fibroblasts, malignancy, as defined by the ability of the fused cells to grow progressively in immunologically compatible hosts, is suppressed. (For reviews see Harris 1986, 1988.) It is also agreed that this suppression is determined by genes located on the chromosomes of the normal fibroblast, so that genes do indeed exist that have the ability to override, or compensate for, the sum total of whatever genetic events might in any particular case be responsible for generating the malignant phenotype. These observations are no longer a laboratory curiosity, but constitute a decisive element in the now substantial body of evidence in support of the view that malignancy may be determined by mutational events that have a genetically recessive character. We know further that a gene or set of genes located on a single normal fibroblast chromosome is both necessary and sufficient to achieve complete suppression of malignancy and that this gene or set of genes acts in a dose-dependent fashion. What we do not know, or at least have rather uncertain ideas about, are the following:

1) Whether any normal cell can suppress malignancy, irrespective of the nature of the tumour cell with which it is fused.

2) Whether the same chromosome is responsible for suppression, irrespective of the normal cell type from which it is derived.

3) Whether suppression is produced by the action of a single gene or by a group of syntenic genes located on one chromosome.
4) Whether suppression is always achieved by the same mechanism.
5) How the suppressive gene or set of genes works. It is these areas of uncertainty that I shall discuss in this paper.

## Can all normal cells suppress malignancy?

While there is now abundant evidence that normal fibroblasts can suppress malignancy when fused with any of a very wide range of different types of tumour cell, comparable experiments with other kinds of normal cells are few. It has been clearly demonstrated (Peehl & Stanbridge 1982) and confirmed (Harris & Bramwell 1987) that normal keratinocytes can suppress malignancy, although, so far, in only a limited range of crosses with epithelial cancer cells. Normal hepatocytes can suppress malignancy when fused with hepatoma cells (Szpirer et al 1988), but the effects of normal hepatocytes when fused with other types of tumour cell have not, as far as I know, yet been explored. Finally, there is some evidence for at least partial suppression of malignancy by normal lymphocytes when these are fused with sarcoma cells (Wiener et al 1974).

Why is this an important question? It is important because the answer to it will define the category of gene that is responsible for the suppression. If, for example, cells from only certain specific differentiation lineages were capable of suppressing malignancy, then genes determining metabolic events common to all cells, such as those governing energy generation, ion transport, DNA replication and so on, would be excluded. But if, on the other hand, any normal diploid cell could suppress, then genes determining specific differentiation pathways would be excluded. Suppression by fibroblasts, apart from a few special cases that involve unusual dosage effects, seems to operate on any malignant tumour cell, irrespective of its histological type or aetiology. This suppression will also operate across species barriers. Although it might appear a simple thing to fuse tumour cells to other types of normal cell and then examine the outcome, problems arise because of preferential elimination from the hybrids of the chromosomes donated by the normal parent cell. For example, when normal mouse lymphocytes are fused with mouse epithelial carcinoma cells or mouse melanoma cells, the hybrids, examined as soon as possible after isolation, show wholesale elimination of the lymphocyte chromosomes. In order to test the ability of the lymphocyte genome to suppress malignancy, special selective techniques are required to ensure retention in the hybrid cell of the particular lymphocyte chromosomes that are of interest. I know of two laboratories that are at present engaged in exploring this question, but because of technical complexities of the kind that I have described, it may be a little time before the outcome is known.

Curiously enough, no systematic karyological analysis by modern methods has yet been made of the so-called hybridomas, hybrids between myelomas and normal lymphocytes.

**What chromosomes are responsible for suppression and how many?**

Shortly after the suppression of malignancy in hybrid cells was first discovered, the suggestion was made that a single normal chromosome might be all that is required to achieve this effect (Harris et al 1969). At least for hybrids between malignant cells and normal fibroblasts, it is now clear that this is indeed the case. Detailed studies extending over many years have provided strong evidence that in mouse crosses, the reappearance of malignancy in hybrids in which it was initially suppressed is determined by the loss from the hybrid cell of the chromosome 4 derived from the fibroblast (Jonasson et al 1977, Evans et al 1982, Copeman & Harris 1988). By exploiting mouse strains in which part of chromosome 4 was translocated to the X chromosome, it was possible to localize the operative region to the upper half of chromosome 4. A deletion that occurred spontaneously within the translocated region permitted further localization to the lower part of the upper half of this chromosome. An intensive effort is now being made by a variety of techniques to map the operative locus more precisely and, if possible, isolate the gene or genes responsible.

Comparable results have been obtained by Levan et al (1988) in hybrids between mouse hepatoma cells and normal rat fibroblasts. A region within rat chromosome 5 was found to be responsible for the suppression of anchorage-independent growth *in vitro*, and examination of linked markers within this region showed that it was homologous with the operative region on mouse chromosome 4. Stoler & Bouck (1985) found that in hybrids between malignant hamster cells and normal human fibroblasts, the suppressive effect was exerted by genes that mapped to human chromosome 1. The presence of linked markers again indicated homology with mouse chromosome 4. It is difficult to avoid the conclusion that the same locus is responsible for the suppression of malignancy in a wide range of different tumours and in different animal species. However, when hepatoma cells were fused with normal rat hepatocytes, no evidence was found of any suppressive effect exerted by rat chromosome 5; in this combination it appeared that the suppression was determined by rat chromosome 8 (Szpirer et al 1988). If this observation is confirmed by further analysis, we shall be able to conclude first, that suppression of malignancy may be achieved by more than one mechanism, and second, that the mechanism of suppression does depend in some way on the type of differentiation that the normal cell shows. I shall have more to say about both of these points.

The extensive cytogenetic studies by Stanbridge et al (1981) and by

Kaelbling & Klinger (1986) on hybrids between human uterine carcinoma cells and normal human fibroblasts concur in finding that the suppression is determined by the normal human chromosome 11. These authors do not, however, see a strong influence exerted by the normal chromosome 1. Their results thus differ from those of Stoler & Bouck (1985). This difference might be due to the fact that Stanbridge et al and Kaelbling & Klinger were scoring suppression of tumour growth *in vivo*, whereas Stoler & Bouck were scoring suppression of anchorage-independent growth *in vitro*. However, the studies implicating mouse chromosome 4 also dealt with the suppression of tumour growth *in vivo*, so that, unless there is a region of homology between the operative locus on mouse chromosome 4 and some gene or genes on human chromosome 11, we again have evidence of suppression being effected by different mechanisms. In the case of human chromosome 11, it has been demonstrated by the technique of monochromosome transfer that the insertion into a tumour cell of a single human chromosome 11 derived from a normal fibroblast is enough to suppress completely the ability of the malignant cell to form a progressive tumour in an animal (Saxon et al 1986). Chromosome 11 can therefore achieve suppression of malignancy without having to interact with any other chromosome in the genome of the normal fibroblast from which it was derived. Similar monochrome transfer experiments with human chromosome 1 are in progress (E.J. Stanbridge, personal communication).

**How many genes?**

Nothing argues against the idea that the locus responsible for the suppression of malignancy is a single gene; but we have no direct evidence that this is so. The most exact localization that we have at present is the deletion in the upper half of mouse chromosome 4, but this is a cytologically visible deletion and involves an amount of DNA that could encompass a large number of genes. An examination of hybrids between mouse melanoma cells and normal human lymphocytes has shown that suppression of malignancy can be achieved by an amount of normal human DNA corresponding to about $3 \times 10^5$ base pairs (Rodgers 1979), but this still leaves room for more than one gene. We must await further progress in the analysis of the operative chromosome region for the resolution of this question.

**What is the mechanism of suppression?**

The evidence at present available leads me to suppose that there are at least two mechanisms of suppression, although they may not be unconnected. Only one of these is illuminated by biochemical evidence. In the case of hybrids between tumour cells and normal fibroblasts, it is clear that the suppression of malignancy is linked to the execution of the terminal differentiation programme of the fibroblast. When hybrids in which malignancy is suppressed are injected into the animal, the cells undergo morphological and physiological changes that exactly reflect the changes that normal fibroblasts undergo under the same conditions (Peehl & Stanbridge 1982, Harris 1985). The hybrids synthesize a copious collagenous extracellular matrix, become aligned into more or less orderly arrays, assume a progressively more elongated fibrocytic morphology and cease to multiply. Malignant segregants that arise from these hybrids as a consequence of having lost the chromosomes 4 derived from the fibroblast do not synthesize any stainable collagenous extracellular matrix, do not undergo any of the characteristic fibrocytic morphological changes and continue to multiply.

A study has recently been made of the extracellular matrix components produced *in vitro* by hybrids between mouse melanoma cells and normal mouse fibroblasts (Copeman & Harris 1988). The melanoma cells synthesize types IV and V procollagen but very little, if any, type I. They also secrete a good deal of protease activity into the medium. Normal fibroblasts, on the other hand, synthesize mainly type I procollagen, but very little type V, and they secrete negligible amounts of protease activity. The hybrids in which malignancy is suppressed, like the normal fibroblasts, synthesize large amounts of type I procollagen, very little type V and secrete very little protease activity. Malignant segregants derived from these hybrids synthesize about one tenth as much type I procollagen as the hybrids in which malignancy is suppressed; but, like the melanoma cells, they synthesize type V procollagen and secrete protease activity. Fusion of the melanoma cell with the normal fibroblast thus suppresses the production of the extracellular matrix components characteristic of the melanoma cell and imposes the production of components characteristic of the fibroblast. When the chromosomes 4 derived from the fibroblast are lost, the fibroblastic extracellular matrix ceases to be produced and elements of the melanoma matrix reappear. These changes in the extracellular matrix produced by the hybrid cells *in vitro* permit one to predict accurately whether or not the cells will produce a progressive tumour when injected into the animal. All these observations are consistent with the idea that there is a gene or set of genes on the fibroblast chromosome 4 that determines both the execution of the differentiation programme of the fibroblast and the suppression of malignancy.

Why should the fibroblast differentiation programme suppress malignancy, when the melanoma differentiation programme does not? My guess is that the differentiation programme of the normal fibroblast is intact and goes to completion, whereas the melanoma programme is incomplete. The completion of the fibroblast differentiation programme *in vivo* produces an immature scar — elongated, non-multiplying cells embedded in a collagenous matrix; and this is essentially what we see in the lesion produced by hybrids in which malignancy is suppressed. When the differentiation programme of normal melanocytes goes to completion, the cells undergo characteristic morphological changes, become heavily pigmented and cease to multiply; but malignant melanoma cells, although they may show some of these features to a variable degree, do not complete the full programme. My general thesis therefore is that malignant cells continue to multiply progressively *in vivo* because they are unable to complete their normal pattern of differentiation, and that malignancy is suppressed in hybrid cells because a complete programme of differentiation is substituted for an incomplete one.

If this thesis is correct, it should in principle be possible to induce progressive growth *in vivo* by blocking a differentiation programme in some other way. On the assumption that, whatever complexities might be involved in the production of a collagenous extracellular matrix, a normal matrix could not be formed if fibronectin production were completely blocked, Diana Donahoe and I have been exploring the possibility of blocking fibronectin production by introducing into the cell cDNA elements that specify antisense messages complementary to parts of the fibronectin gene. Vectors were constructed in which the fibronectin antisense cDNA, with an appropriate promoter and polyadenylation signal, was ligated to selectable markers. These vectors were then introduced by electroporation into melanoma × normal fibroblast hybrids which showed suppression of malignancy and which produced a copious collagenous extracellular matrix. Several transfected clones were isolated in appropriate selective medium and were found by Southern blotting to have incorporated the fibronectin antisense DNA. Northern blotting revealed that the incorporated antisense DNA was transcribed. The majority of these clones showed only trace amounts of fibronectin on their surfaces and were found to be highly malignant when injected into appropriate syngeneic hosts. In some cases, the take incidences and the latent periods before the appearance of tumours were comparable to those seen with the melanoma parent cell itself. Immunohistological examination of the tumours showed that they contained very little fibronectin, and they produced no stainable extracellular matrix. Karyological examination of these transfected clones showed that they were not composed simply of cells that for some obscure reason had eliminated the chromosomes 4 derived from the normal fibroblast. Indeed, the results suggested that if the ability of the hybrid cell to produce a normal collagenous extracellular matrix is impaired in some other way, the selective pressure in favour of cells that have lost the normal fibroblast chromosomes 4 is removed. Malignant clones that failed to produce fibronectin were not obtained in control experiments where the corresponding fibronectin cDNA was introduced into the cells in the sense orientation. These observations, of course, need confirmation and further exploration but they encourage the belief that it may be possible to induce the reappearance of malignancy in hybrid cells by specific inhibition of a single defined step in a differentiation pathway.

The general principles that have been extracted from the study of hybrids between tumour cells and fibroblasts apply also to hybrids between tumour cells and normal keratinocytes. In this case, the suppression of malignancy is associated with the imposition on the hybrid cell of the terminal differentiation programme of the keratinocyte. When injected into animals the hybrid cells undergo typical squamous differentiation, form keratin nests and cease to multiply (Peehl & Stanbridge 1982). In the malignant segregants derived from these hybrids, squamous differentiation fails to occur and elements of the differentiated phenotype of the malignant parent cell reappear. It has been shown further that hybrids in which malignancy is suppressed synthesize involucrin, a key component in the terminal differentiation pathway of keratinocytes, whereas malignant segregants derived from these hybrids have lost the ability to synthesize involucrin (Harris & Bramwell 1987). Once again we see the suppression of malignancy associated with the substitution of a complete differentiation programme for an incomplete one.

But why should this substitution occur? At this point I am driven to consider the question of gene dosage in the suppression of malignancy. It is a cardinal principle of cell biology that a cell committed to one programme of differentiation cannot simultaneously carry out another. So when two cells from different (often obviously incompatible) differentiation lineages are fused together, the hybrid cell has to make a choice. Why is it that when tumour cells are fused with fibroblasts, it is the fibroblast differentiation programme that generally prevails, even in those rare hybrids which are karyologically complete, that is, retain the entire chromosome sets of both parent cells? One possible explanation is that in selecting for cells that grow on glass or plastic we favour the emergence of clones with a fibroblastic phenotype. Perhaps if we selected for cells that grew in suspension or in agar we might favour the emergence of other phenotypes. There is some evidence in support of this idea, but I doubt that it is the complete explanation.

We know that the suppression of malignancy, and the imposition of a fibroblastic differentiation programme, by the gene or set of genes on mouse fibroblast chromosome 4 is dosage dependent. If in the hybrid one contrives to increase the number of chromosomes 4 derived from the tumour parent cell relative to the number derived from the normal fibroblast, then suppression of malignancy is relieved and fibroblastic differentiation does not take place (Evans et al 1982). It is thus apparent that the chromosomes 4 derived from the tumour cell are producing something that competes with the product(s) of fibroblast chromosomes 4. We can assume that, in the case of the normal fibroblast, whatever genes are involved are present in two copies. But we know nothing about the state of the relevant genes on the chromosome 4 in the tumour cell. Since it is now clear that hemizygosity is an extremely common occurrence in tumour cells, it is possible that only one functional copy of the gene or genes of interest might be present in the tumour cell. In that case, the predominance of the fibroblast phenotype might be explicable in terms of the numerical preponderance of a gene or set of genes, located on chromosome 4, that governs the fibroblastic differentiation programme in a pleiotropic manner.

The other mechanism that might be responsible for the suppression of malignancy is easily explicable in formal genetic terms, although its existence has yet to be directly demonstrated. It is becoming increasingly obvious that homozygous recessive mutations are involved in the genesis of a very wide range of malignancies, both juvenile and adult. Indeed, the rate of discovery of homozygous recessives in malignant tumours leads me to suppose that they will be found in most, if not all, malignant tumours. Whether there are any cellular genes, mutated or otherwise, that determine malignancy in a genetically dominant fashion remains, in my view, an open question. If we assume that homozygous recessive mutations are responsible for the malignancy of the tumours that have been used in cell fusion experiments, then one way in which malignancy could be suppressed would be by complementation of these mutations by wild-type alleles. I suspect that this is what might be happening when one suppresses malignancy by the introduction into the tumour cell of a single normal chromosome. Since most differentiation programmes involve the regulation of a sizeable number of proteins whose genes may map to several different chromosomes, it seems unlikely that a complete differentiation programme would be transferred by a single chromosome, unless that chromosome were able to re-programme the genome of the recipient cell. But this does not seem to happen, for the introduction of chromosome 11 from a human fibroblast into an epithelial tumour cell suppresses malignancy without inducing fibroblastic differentiation in the recipient cell (E.J. Stanbridge, personal communication). Moreover, when the malignancy of hepatomas is suppressed by fusion with normal hepatocytes, no new differentiation programme is imposed on the hybrid cells, and, as mentioned earlier, the operative chromosome in this case appears to be different from the one responsible for the suppression of malignancy when these same hepatoma cells are fused with fibroblasts, where a fibroblastic pattern of differentiation is imposed on the hybrids. These two mechanisms could, however, be subsumed in a unifying hypothesis if the homozygous recessive mutations in tumour cells blocked, either directly or indirectly, some critical step in a terminal differentiation programme. In that case, suppression of malignancy in hybrid cells could be achieved either by overcoming this block or by substituting a different, but intact, programme for the defective one.

## Conclusion

The results of two decades of work on the suppression of tumour formation by cell fusion are entirely consonant with the current wave of interest in recessive mutations as determinants of cancer. What is not clear is where, if at all, the heterogeneous collection of genes known as 'oncogenes' fits into the picture. My guess is that if these genes do contribute to the genesis of malignancy in any immediate way, they do so by impairing the function of cellular genes that are involved in the process of differentiation.

## References

Copeman MC, Harris H 1988 The extracellular matrix in hybrids between melanoma cells and normal fibroblasts. J Cell Sci 9:281–286

Evans EP, Burtenshaw MD, Brown BB, Hennion R, Harris H 1982 The analysis of malignancy by cell fusion. IX Re-examination and clarification of the cytogenetic problem. J Cell Sci 56:113–130

Harris H 1985 The suppression of malignancy in hybrid cells: the mechanism. J Cell Sci 79:105–117

Harris H 1986 The genetic analysis of malignancy. J Cell Sci (suppl) 4:431–444

Harris H 1988 The analysis of malignancy by cell fusion: 1988. Cancer Res 48:3302–3306

Harris H, Bramwell ME 1987 The suppression of malignancy by terminal differentiation: evidence from hybrids between tumour cells and keratinocytes. J Cell Sci 87:383–388

Harris H, Miller OJ, Klein G, Worst P, Tachibana T 1969 Suppression of malignancy by cell fusion. Nature (Lond) 223:363–368

Jonasson J, Povey S, Harris H 1977 The analysis of malignancy by cell fusion. VII Cytogenetic analysis of hybrids between malignant and diploid cells and of tumours derived from them. J Cell Sci 24:217–254

Kaelbling M, Klinger HP 1986 Suppression of tumorigenicity in somatic cell hybrids. III Cosegregation of human chromosome 11 of a normal cell and suppression of tumorigenicity in intraspecies hybrids of normal diploid × malignant cells. Cytogenetic Cell Genet 41:65–70

Levan G, Islam MQ, Szpirer J, Szpirer C 1988 A tumour suppressor gene assigned to rat chromosome 5 region q2. Abstract in Tumour Suppressor Genes and Negative Growth Regulation, EMBO/NIEHS/CRC Workshop, Nethybridge, Scotland

Peehl DM, Stanbridge EJ 1982 The role of differentiation in the suppression of tumorigenicity in human hybrid cells. Int J Cancer 30:113–120

Rodgers A 1979 Detection of small amounts of human DNA in human-rodent hybrids. J Cell Sci 38:391–403

Saxon PJ, Srivatsan ES, Stanbridge EJ 1986 Introduction of human chromosome 11 via microcell transfer controls tumorigenic expression of HeLa cells. EMBO (Eur Mol Biol Organ) J 5:3461–3466

Stanbridge EJ, Flandermeyer RR, Daniels D, Nelson-Rees WA 1981 Specific chromosome loss associated with the expression of tumorigenicity in human cell hybrids. Somatic Cell Genet 7:699–712

Stoler A, Bouck N 1985 Identification of a single chromosome in the normal human genome essential for suppression of hamster cell transformation. Proc Natl Acad Sci USA 82:570–574

Szpirer C, Szpirer J, Islam Q, Levan G 1988 Genetic analysis of growth control in hepatoma cell hybrids. Abstract in Tumour Suppressor Genes and Negative Growth Regulation, EMBO/NIEHS/CRC Workshop, Nethybridge, Scotland

Wiener F, Klein G, Harris H 1974 The analysis of malignancy by cell fusion. V Further evidence of the ability of normal diploid cells to suppress malignancy. J Cell Sci 15:177–183

## DISCUSSION

*zur Hausen:* If I understand you correctly, the 'tumour suppressing genes' in, for instance, fibroblast–HeLa cell hybrids would be different from those in keratinocyte–HeLa cell hybrids?

*Harris:* This is a plausible prediction, but I do have some evidence that suggests that regulatory genes governing different modes of differentiation may map to the same chromosome.

*Bodmer:* Is the fibronectin gene in the mouse on chromosome 4?

*Harris:* I don't know where the mouse fibronectin gene maps, but it doesn't matter. The extracellular matrix contains a variety of compounds and they map to different loci. I chose to start with fibronectin, not because I thought it was a key component, but because I thought that if I could knock out fibronectin production, the remaining components wouldn't make a proper extracellular matrix. I will repeat the experiment with other components, but if the genes are very big, as they are in the case of the collagens, it is a problem to get them into a cell without scrambling the DNA.

*Bodmer:* Your crosses are always between different tissue types, so there must be some involvement of the extinction phenomenon. To what extent is the switching on and off of collagen a part of that extinction phenomenon? How different would it be if you crossed your melanoma with a normal melanocyte?

*Harris:* It is not true that all my experiments have been done with disparate histiotypic crosses. I have done, and published, many experiments in which normal fibroblasts have been crossed with sarcomas of various kinds, including fibrosarcomas. With respect to the suppression of malignancy, the results are no different from those obtained with disparate histiotypic crosses.

I do not think this is an appropriate time to enter deeply into a discussion of the phenomenon of 'extinction'. I should merely like to say that, in my view, 'extinction', as Boris Ephrussi envisaged it, namely the suppression in the hybrid of so-called 'luxury functions' that were expressed in the parental cells, does not exist. All that happens is that you substitute one differentiation pattern, usually fibroblastic differentiation, for another.

*Marks:* How strongly do you insist on the association of suppression of tumorigenicity and expression of complete terminal differentiation? There are a number of examples where the suppression of malignancy is associated with only partial expression of the differentiation programme.

*Harris:* I don't insist on it at all. I have chosen to work on the last steps of keratinocyte differentiation a) because the genes have been isolated and are available, and b) because the intermediate steps of most forms of differentiation remain obscure. The important point is that involucrin, cross-linked by transglutaminase, is undoubtedly a tumour suppressor, but it is also a perfectly normal differentiation gene product. But involucrin is only a tumour suppressor when the transglutaminase is present, and *vice versa*. This refers to Al Knudson's question of how you override malignancy. I believe that all malignant cells are blocked at some point in their differentiation.

*Sporn:* I still don't understand the connection between the turn on of differentiation and the suppression of malignancy. What is the mechanistic connection between these two that you are postulating?

*Harris:* In the case of keratinocyte differentiation, it's a very crude mechanistic connection. If you take a cell and put in a gene that makes involucrin, and the cell has another gene that cross-links the involucrin, you make an insoluble cell envelope. There are more subtle ways of suppressing malignancy, but I have taken an extreme case to make a point of principle about what a tumour suppressor gene is.

*Sporn:* But you have just wrapped up the cell and put a brick wall around it.

*Harris:* That is true, but that is exactly what squamous differentiation is. In the fibroblast, you make an extracellular matrix. If you like, you put a wall of collagen around the cell, but, in that case, there has to be an interaction between the collagen and the cell.

*Stanbridge:* You believe that the same locus on chromosome 4 governs differentiation and suppression. What would you predict if you were able to do a monochromosome transfer into the melanoma using the fibroblast chromosome 4? Would there be suppression and/or differentiation, and what form do you think that differentiation would take?

*Harris:* There may or may not be suppression, but my guess is that you would not impose the fibroblast differentiation programme. My reason is the following. It appears that there is a locus on chromosome 4 that governs the whole, or a large part, of the fibroblast differentiation programme in a pleiotropic fashion. There are in other systems good examples of this kind of pleiotropic regulation. A set of genes involved in a particular pathway have nucleotide sequences in common. In some cases, it has been shown that some other gene makes a product that recognizes these common sequences in the different genes and switches them on or off coordinately. My guess is that the regulatory locus on chromosome 4 would switch on the fibroblast differentiation genes within a fibroblast. To do this in, say, a colonic carcinoma cell, the regulatory locus would have to achieve the complete re-programming of the genome of that particular cell.

The general experience in hybrid cells is that the main differentiation lineages are stably maintained in the hybrid. We did an experiment many years ago where we took a lymphoma, which looked much like a lymphocyte, and crossed it with a diploid fibroblast. The hybrids looked like fibrosarcoma cells. They maintained that morphological phenotype in culture for months. When the cells were injected into an animal, segregants emerged in which the whole of the lymphocyte differentiation programme was restored. So although visible lymphocyte differentiation was abrogated in culture, the lymphocyte programme was nonetheless maintained in a cryptic form.

*Stanbridge:* Would you predict suppression by introduction of the fibroblast chromosome 4 into the melanoma? If you do predict suppression, would you predict any form of differentiation, such as melanocyte-type differentiation?

*Harris:* I would not predict that with chromosome 4 alone you would necessarily change the whole of one differentiated programme for another. Whether you would suppress malignancy or not is unpredictable. If you put in the whole of chromosome 4, you put in a lot of genes that may alter all sorts of things in a subtle way, and you might end up suppressing malignancy, but not by imposing the differentiation pattern of the cell from which the chromosome 4 was derived.

*Stanbridge:* So when you say that the same locus on chromosome 4 governs differentiation and suppression, that is your guess when other chromosomes of the fibroblast are present in that hybrid.

*Harris:* Yes.

*Lau:* Professor Harris, a further prediction from your model would be that after the hybrids have acquired the fibroblast differentiation pattern, if you passage them long enough, because of progressive loss of chromosomes other than chromosome 4, it would come to a point when the hybrid could no longer maintain the fibroblast differentiation pattern but would revert back to the melanoma differentiation programme. It would still be non-tumorigenic because of the presence of chromosome 4. Have you ever actually observed that?

*Harris:* That would have to be examined in a systematic way. I don't have the answer.

*Bodmer:* If you put chromosome 4 from the fibroblast into a fibrosarcoma, I predict that you would block the tumorigenicity and induce fibroblast differentiation.

*Stanbridge:* I don't believe it is quite that simple. In the HeLa–fibroblast human hybrids, introduction of a normal chromosome 11 into a HeLa × fibroblast tumorigenic segregant suppresses the tumorigenic phenotype, apparently by inducing differentiation into fibroblastoid tissue. When we introduced the same chromosome 11 into a fibrosarcoma by microcell transfer, there was no effect at all.

*Harris:* Try chromosome 1. There is a very important generalization underlying our discussions but not yet explicitly stated. We are all talking as if a particular chromosome from a fibroblast is *genetically* different from the same chromosome derived from some other cell type. That means that stable forms of differentiation are imprinted on particular chromosomes. In terms of the theory of differentiation, you couldn't say anything more important. I have more or less believed this for a long time but your experiments, Eric (Stanbridge, this volume), will finally decide whether a chromosome derived from a cell of one differentiation lineage is genetically different from the same chromosome derived from a cell of a different lineage.

*Bodmer:* You can account for that, for example, by arguing about the degree of methylation.

*Harris:* That's an obvious possibility, of course.

*Klein:* I was very pleased about the connection you made between eclipse of differentiation and eclipse of malignancy. If fits beautifully with your and Eric's work on hybrids. Thinking of all those hybrids we tested in the early days, your concept fits very nicely with most of them. For example, the normal fibroblast is suppressive whatever the partner is. But one problem is that your hypothesis would predict lymphoma-like differentiation in the sarcoma–normal lymphocyte hybrid, but that was not the case. These hybrids were totally fibroblast-like and yet they were suppressed. How can that be accommodated in your general model?

An even more difficult problem is our single complementation hybrid between the lymphoma YACIR and the sarcoma MSWBS. These are two malignant tumours and the hybrid showed reduced malignancy. Malignant segregants had usually a sarcoma-like morphology, but occasionally we also saw reappearance of the lymphoma- like morphology.

*Harris:* I have for a long time been dissatisfied with the depth of the studies that we did in those early days on the lymphocyte hybrids. Clare Madin and I are now systematically re-making all those lymphocyte hybrids and others. Our earlier hybrids were examined before chromosome banding was discovered, and there was no decent karyology, so we don't really know what went on.

I can describe our results so far. If we start with a mouse melanoma and cross it with a normal mouse lymphocyte, there is extreme segregation of the normal lymphocyte chromosomes. We haven't yet been able to get a hybrid, for example, that has retained the lymphocyte chromosome 4. If we cross the melanoma with YACIR, which is a malignant mouse lymphoma, then we obtain lots of hybrids that are pretty complete and contain the YACIR chromosome 4. We are also making crosses with lymphocytes from the T37H mouse. The T37H mouse has a translocation of the relevant part of chromosome 4 onto the X chromosome. Because of the translocation, we have a selectable marker, so we are forcing the retention of the relevant part of chromosome 4. One problem is that we don't have good markers that would enable me to say that the differentiation programme of the lymphocyte has gone to completion. The completion of the lymphocyte differentiation programme is a non-multiplying memory cell, and I don't know what that phenotype looks like on a fibroblast background.

Is anybody interested in why the fibroblast differentiation programme is imposed? Since there are two differentiation programmes, why does the fibroblast programme usually predominate? I can think of two possible explanations. One is that, in general, we select our hybrids for their ability to grow attached to plastic or glass surfaces, and that selects pretty strongly against lots of other phenotypes. If we selected only those hybrids that grew in suspension, it would not surprise me if other phenotypes emerged.

But I think there may be a more fundamental reason. We know that when we fuse with a normal fibroblast, we introduce into the hybrid two functional alleles of the operative locus determining fibroblast differentiation. However, we do not know how many functional alleles there might be determining the differentiation pattern of the malignant cell we are fusing the fibroblast with. Since hemizygosity is extremely common in malignant cells, they may only contain one such functional allele, so that the system is loaded in favour of the fibroblast.

*Wyke:* I am a bit puzzled by your requirement for the explanation. I understood you to say that the *sine qua non* of the malignancy of the melanoma is that it cannot execute a differentiation programme.

*Harris:* It can't *complete* its differentiation programme but it does have some markers of differentiation.

*Wyke:* An even simpler explanation is that the cell has no choice but to execute the fibroblast differentiation programme. The other possibility is that melanoma differentiation cannot be carried out, because of hybrid extinction of differentiated functions, the mechanisms for which are not understood.

*Harris:* As I have already explained, I don't believe differentiation programmes are extinguished in any sense other than that they are replaced by other differentiation programmes. There is no such thing as a cell without a differentiation programme. When you say that differentiated functions are extinguished, for example, pigmentation, all that you are actually observing is the substitution of a fibroblast differentiation programme for the melanoma differentiation programme. The question is, why the substitution?

*Croce:* You avoid the problem of extinction by saying that the hybrid cell might substitute or partially or completely replace the function of the tumorigenic parent. What about the somatic cell hybrid between a malignant cell and a normal cell which belong to the same lineage and are at more or less the same stage of differentiation?

*Harris:* If you mean hybrids between cells of the lymphoid series, I have said my piece in print about those (Harris 1988). Despite the fact that the so-called hybridomas are so widely used, there has been no serious karyology done on them. The chromosome constitutions of the standard myelomas that are used in those crosses are grotesque: the cells are usually subtetrapoloid, full of translocations, breaks and other aberrations. I know that when you fuse a myeloma cell with a normal lymphocyte there is extreme segregation of the normal lymphocyte chromosomes. Whether a normal lymphocyte can suppress the malignancy of a myeloma remains to be shown. When you put hybridomas into the peritoneal cavity soon after making them, the cells often don't grow straight away. There is a lag period of up to 3–4 weeks. If you look at the chromosomes of the cells, it is quite clear that you are selecting a subpopulation that grows in the peritoneal cavity. So if you are going to argue about whether a normal lymphocyte can suppress the malignancy of a melanoma, let me assure you that a) it is a very difficult experiment to do properly and it's going to take a lot of your time, because of the extreme segregation of the normal lymphocyte chromosomes, and b) the proper experiment has never been done.

*Croce:* There was a paper describing somatic cell hybrids between mouse myeloma and lymphocytes, claiming that those hybrids were all tumorigenic.

*Harris:* That was by Dario Giacomoni (1979) and the work was done in my lab. Most of those hybrids were malignant, but there was no analytical karyology. Only gross chromosome counts were made.

*Klein:* I agree with Henry's view that the karyology is not convincing, but I have a lingering thought that there might be an exception. The normal lymphocyte is a resting cell that can be activated to divide by a variety of mitogens. If you fuse it with a plasmacytoma cell programmed for division, it is probably just activated and does not suppress. We have seen a similar situation in lymphoma–lymphocyte hybrids in the Spira study. Normal lymphocytes didn't suppress JIKAUT, whereas fibroblasts and macrophages did.

*Benedict:* There is another fact that seems to be important in this hybrid story. 95% or more of the hybrids that we make never survive for more than a few cell divisions, i.e. not to the point where we can grow enough cells to investigate this question of suppression of tumorigenicity. So I believe there is another gene, which might be called a senescence gene, that takes precedence over any of the genes that we have discussed today.

*Harris:* I wouldn't argue against that. To do the sort of experiments I have been talking about, we have to have clones that grow.

## References

Harris H 1988 The analysis of malignancy by cell fusion: the position in 1988. Cancer Res 48:3302–3306

Giacomoni D 1979 Tumorigenicity and intracisternal A-particle expression of hybrids between murine myeloma and lymphocytes. Cancer Res 39:4481–4484

Stanbridge EJ 1989 A genetic basis for tumour suppression. In: Genetic analysis of tumour suppression. Wiley, Chichester (Ciba Found Symp 142) p 149–165

# General discussion V

## Collagen-induced differentiation in a colorectal cancer cell line

*Bodmer:* I would like to tell you about a model we have for differentiation in the colorectal system. We have been doing a genetic and biochemical analysis of differentiation in colorectal epithelium. The rationale is that differentiation is one component of the malignant progression that we can try to control and understand.

P Richman and M Pignatelli in my lab screened a series of colorectal carcinoma-derived cell lines for conditions that would lead to induction of differentiation. A particular cell line, called SW1222, when put on a type 1 collagen gel forms structures that look like part of a colonic crypt. A pathologist would say that was a low grade tumour that was differentiated and had a relatively good prognosis. That differentiation opposes growth and high malignancy; if you try to subculture those structures, you can't.

We felt that it was important, when studying an aspect of differentiation, to do the genetics in an homologous system. So we obtained a mouse colorectal carcinoma-derived cell line from LM Franks to give us the correct background in which to look at these processes. We haven't, however, done the controls of what happens with other sorts of hybrids. That mouse cell line, like most colorectal carcinoma cell lines, does not differentiate in this way when put onto a collagen gel. In the somatic cell hybrids between SW1222 and the mouse line the ability to differentiate co-segregates with chromosome 15.

The most likely explanation for this process was that it could be a direct induction by attachment to collagen. Piersbacher and Ruoslahti (Dedhar et al 1987) have shown that there are specific receptors for type I collagen, which involve the RGD peptide motif, and the receptors for collagen type 1 are specifically blocked by the peptide RGDT. Using a simple quantitative measure of the extent of differentiation, we showed that differentiation of the SW1222 cell line is significantly blocked by the peptide. A control peptide containing glutamic acid instead of the aspartic acid does not block. So the induction of this differentiation is specifically blocked by blocking of binding to collagen. We have also been able to show that this induction correlates directly with binding to collagen. Thus, we believe we are looking at specific induction of differentiation due to attachment to type I collagen by a receptor that has the RGDT binding property, and that this is what we are mapping in the somatic cell hybrids (Pignatelli & Bodmer 1988).

*Marshall:* Is that concentration of peptide about the same range as in other RGD blocking experiments?

*Bodmer:* Yes, I think so. In all the cell lines we have looked at, there is a complete correlation between binding to type I collagen and blocking by the RGDT peptide and any differentiation of this sort. Our overall results suggest that fibronectin and other extracellular matrix components are not involved.

If you put one of the hybrids that carries human chromosome 15 into a nude mouse, you see the differentiated structures. Although we haven't convincingly shown it yet, it seems most likely that you don't get a tumour, because these structures don't grow *in vitro*. Hybrids that don't have chromosome 15 don't form these structures. This shows that the differentiation process is induced in the nude mouse, presumably by attachment of the hybrid cells to the mouse extracellular matrix components. It also shows a lack of species specificity of the human versus the mouse collagen type I receptor. We now need to do more detailed genetic and biochemical studies to show that this induction of differentiation is due to the presence of a type I collagen receptor.

*Sporn:* Sir Walter, do you want to say anything about the effects of TGF-β on these cells? The genes for the receptors for all these matrix substances are now known to be controlled by TGF-β.

*Bodmer:* Massimo Pignatelli has been adding TGF-β to these systems and there are some intriguing results which we don't yet fully understand. In the cell line SW1222, there is a much stronger induction of differentiation in the presence of TGF-β. There is a better polarization of the cells, so that morphologically they look much more like the cells in the crypt. Interestingly, if you add TGF-β to the somatic cell hybrid that carries chromosome 15 and one or two others, you don't see an effect. The impression is that the TGF-β effect on differentiation is probably a combination of an effect on the surface receptor and other things that are influencing the pattern of differentiation.

*Croce:* Do the hybrids that contain human chromosome 15 carry it in 100% of the cells? Otherwise you would expect to see a sort of a mixture.

*Bodmer:* You do see a mixture, but you don't see 100% of the parent SW1222 cell making these structures and we don't know why that is. When the structures are made we can tell whether chromosome 15 is there by staining for $\beta_2$-microglobulin and we can get reasonably uniform cell populations by selection using a fluorescence-activated cell sorter with an anti-$\beta_2$ antibody.

*Stanbridge:* Can you enrich for the SW1222 cells that form the glandular structures by taking those structures and reculturing them *in vitro*?

*Bodmer:* No, because those structures are post-mitotic.

*Marshall:* The model you put forward is that chromosome 15 is coding for a collagen receptor. An alternative is that it is due to differences in expression of collagenase. Collagenase is a protein kinase C-activated gene, and SW1222 may be one of those unusual colon carcinomas that doesn't have an activated *ras* gene and for that reason does not express collagenase.

*Bodmer:* We can block the binding with RGDT so there has to be a receptor there. We can also show labelled collagen binding to the hybrid that has chromosome 15 and not to others.

*Marshall:* But you haven't looked at collagenase expression?

*Bodmer:* We haven't, but we will do. The most obvious thing is to clone the receptor gene or get clones to that collagen type I receptor.

## References

Dedhar S, Ruoslahti E, Pierschbacher MD 1987 A cell surface receptor complex for collagen type I recognizes the Argo-Gly-Asp sequence. J Cell Biol 104:585–592

Pignatelli M, Bodmer WF 1988 Genetics and biochemistry of collagen binding-triggered glandular diffentiation in a human colon carcinoma cell line. Proc Natl Acad Sci USA 85:5561–5565

# Cell differentiation and tumour suppression

Leo Sachs

*Department of Genetics, Weizmann Institute of Science, Rehovot 76100, Israel*

*Abstract.* Identification of growth and differentiation-inducing proteins and how they interact in normal development has made it possible to identify the molecular basis of normal development and the mechanisms that uncouple growth and differentiation so as to produce malignant cells. When normal cells have been changed into cancer cells, the malignant phenotype can again be suppressed. Results on the molecular control of growth and differentiation in normal myeloid haemopoietic cells, changes in the normal developmental programme in myeloid leukaemia, and the suppression of malignancy in myeloid leukaemia and sarcomas have shown that (1) malignancy can be suppressed either with or without genetic changes in the malignant cells, (2) suppression of malignancy by inducing differentiation does not have to restore all the normal controls, and (3) genetic abnormalities which give rise to malignancy can be bypassed and their effects nullified by inducing differentiation which stops cells from multiplying.

*1989 Genetic analysis of tumour suppression. Wiley, Chichester (Ciba Foundation Symposium 142) p 217–233*

The multiplication and differentiation of normal cells are controlled by different regulatory molecules. These regulators have to interact to achieve the correct balance between cell multiplication and differentiation during embryogenesis and the normal functioning of the adult individual. The origin and progression of malignancy result from genetic changes that uncouple the balance between multiplication and differentiation so that there are too many growing cells. This uncoupling can occur in various ways (Sachs 1980, 1987a). When cells have become malignant, how can malignancy be suppressed so as to revert them back to non-malignant cells? Malignant cells can have different abnormalities in the controls for multiplication and differentiation. Do all the abnormalities have to be corrected, or can they be bypassed in order to suppress malignancy? In this paper I mainly discuss results with normal and leukaemic myeloid haemopoietic cells and sarcomas that have been used as model systems to answer these questions.

## Normal growth- and differentiation-inducing proteins

An understanding of the mechanisms that control multiplication (growth) and differentiation of normal cells would seem to be an essential requirement for

**TABLE 1 Normal myeloid haemopoietic growth and differentiation-inducing proteins**

| | | | Induction of differentiation | |
|---|---|---|---|---|
| | *Nomenclature*[b] | *Differentiated cell type* | *Direct* | *Indirect*[a] |
| Growth inducers | MGI-1M = M-CSF = CSF-1 | Macrophages | − | + |
| | MGI-1G = G-CSF | Granulocytes | − | + |
| | MGI-1GM = GM-CSF | Macrophages and granulocytes | − | + |
| | IL-3 | Macrophages, granulocytes and others | − | + |
| Differentiation inducers | MGI-2 = DF = IL-6 | Macrophages and granulocytes | + | − |

[a] Growth-inducing protein activates production of differentiation-inducing protein. [b] See text for definitions.

elucidating the origin and reversibility of malignancy. The establishment of a cell culture system for the clonal development of normal haemopoietic cells (Ginsburg & Sachs 1963, Pluznik & Sachs 1965, Bradley & Metcalf 1966) and discovery of the regulators of this clonal development in cell culture supernatants (Pluznik & Sachs 1966, Ichikawa et al 1966) have led to the identification of a family of different myeloid cell growth and differentiation-inducing proteins. These include four different myeloid cell growth-inducing proteins called macrophage and granulocyte inducers-type 1 (MGI-1), or colony-stimulating factors (CSF) and interleukin 3 (IL-3) (Table 1) (see Sachs 1982, 1987b,c, Metcalf 1985). Cloning of the genes from mouse and human for these four growth inducers has shown that these genes are unrelated in their nucleotide sequences (see Clark & Kamen 1987).

How do normal myeloid precursor cells induced to multiply by these growth inducers develop into clones that contain mature differentiated cells which stop multiplying when they terminally differentiate? It appears unlikely that a growth inducer which induces cell multiplication is also a differentiation inducer whose action includes the stopping of cell multiplication in mature cells. Indeed, proteins that induce myeloid cell differentiation but not growth have been identified and are called macrophage and granulocyte inducers-type 2 (MGI-2) or differentiation factors (see Sachs 1980, 1982, Tomida et al 1984, Olsson et al 1984). Each of the myeloid growth and differentiation-inducing proteins has a different cell surface receptor (Lotem & Sachs 1986, 1987). Studies on amino acid sequence, neutralization by monoclonal antibody (Shabo & Sachs 1988) and myeloid cell differentiation-inducing activity of recombinant protein have shown that MGI-2 is interleukin 6 (IL-6) (Shabo et al 1988).

Experiments with normal myeloid precursor cells have shown that in these cells any of the four growth-inducing proteins prevents cell death, and induces cell multiplication and production of differentiation-inducing protein (Sachs 1980, 1982, 1987c). The myeloid differentiation inducers activate differentiation directly, whereas the myeloid growth inducers activate differentiation indirectly by inducing the production of differentiation inducer (Table 1, Fig. 1). This induction of differentiation inducer by growth inducer is thus an effective mechanism to couple normal growth and differentiation, a coupling mechanism that may also apply to other cell types. Differences in the time of activation of the differentiation inducer would produce differences in the amount of cell multiplication that occurs before differentiation. A differentiation inducer can also switch on production of a growth inducer in myeloid cells (Lotem & Sachs 1986). The studies on myeloid cells have thus shown that there are different proteins that participate in the developmental programme, growth inducers and differentiation inducers, and that there is a cascade of interactions between growth inducers and differentiation inducers in development.

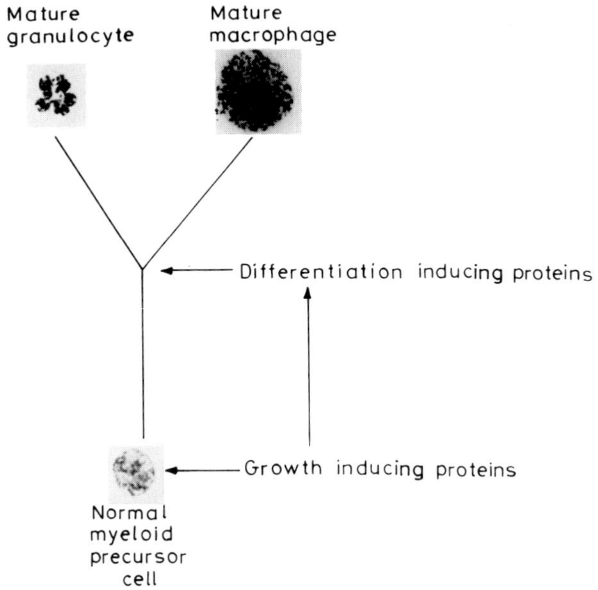


FIG. 1. Growth of normal myeloid haemopoietic precursor cells and their differentiation to macrophages or granulocytes are induced by different proteins: growth-inducing proteins and differentiation-inducing proteins. The growth-inducing proteins induce cell viability and multiplication of normal precursors and production in these cells of differentiation-inducing proteins. This induction of differentiation inducer by growth inducer provides a mechanism to couple the multiplication of normal precursors and their differentiation.

## Growth-inducing proteins and malignancy

The normal myeloid growth inducers can be produced by various cell types (Sachs 1974). However, these growth inducers are not made by the normal myeloid precursors, which require the production of growth-inducing protein by other cell types for cell viability and growth. When cells become malignant they have escaped some normal control, which can be associated with changes from an induced to a constitutive expression of certain genes (see Sachs 1980, 1982, 1987b). In myeloid leukaemic cells, different clones of malignant cells have been identified which have shown the various types of changes that can occur in the normal response to growth inducers. There are different leukaemic clones that (1) need less or have become independent of normal growth inducer for growth, (2) constitutively produce their own growth inducer, or (3) are blocked in the ability of growth inducer to induce production of differentiation inducer (see Sachs 1978, 1980, 1982, 1987b). Thus, one way in which malignancy can originate is by changes in the production of, or response to, a normal growth inducer which gives a growth advantage to the leukaemic over the normal cells.

Independence from normal growth inducer or constitutive production of their own growth inducer can also explain the survival and growth of metastasizing malignant cells in places in the body where the growth inducer required for the survival of normal cells is not present. In cells that are malignant but may still need some growth inducer, the organ preference of metastasis could be due to production of the required growth inducer in the organ where the metastasis occurs.

**Induction of differentiation in malignant cells**

The different types of myeloid leukaemic cells include clones that have changed their normal requirement for growth inducer and in which growth inducer no longer switches on production of differentiation inducer. These clones, which are called $D^+$ clones (D for differentiation) (Fibach et al 1973), can be induced to differentiate to mature macrophages or granulocytes via the normal sequence of gene expression by incubating the cells with normal myeloid differentiation-inducing protein (Fig. 2) (see Sachs 1974, 1978, 1980, 1982, 1987b). The mature cells, which can be formed from all the cells of a leukaemic clone, then stop multiplying like normal mature cells and are no longer malignant *in vivo* (see Lotem & Sachs 1981, Sachs 1982, 1987b,c). Studies in animals have shown that differentiation of $D^+$ myeloid leukaemic cells to mature non-dividing cells can also be induced *in vivo* (see Sachs 1978, 1982, 1987c). These leukaemias grow progressively when there are too many leukaemic cells for the normal amount of differentiation inducer in the body. The development of leukaemia in mice with these $D^+$ leukaemic cells can be inhibited by increasing the amount of normal differentiation-inducing protein, either by injecting it or by injecting a compound that increases its production by cells in the body (see Lotem & Sachs 1981, Sachs 1982, 1987b). After injection of these myeloid leukaemic cells into fetuses, the leukaemic cells can participate in haemopoietic cell differentiation in apparently healthy adult animals (Gootwine et al 1982).

There are other $D^+$ clones that can be induced to differentiate by incubating the leukaemic cells with a normal myeloid growth-inducing protein (Table 2). In these leukaemic clones the growth inducers can presumably still induce production of an appropriate differentiation inducer. Not all clones respond to the same protein even though they may have the appropriate receptors. The results show that most, and probably all, of the physiological regulatory proteins of normal myeloid haemopoietic cell development can regulate differentiation of $D^+$ myeloid leukaemic cells (Table 2) (Lotem & Sachs 1988). In one of these clones (Table 2, WEHI-3B) the genetic abnormalities include a rearrangement of Hox 2.4 homeobox gene (Blatt et al 1987, 1988) and in other clones there is a deletion of one copy of Hox 4.1 (Blatt & Sachs 1988).

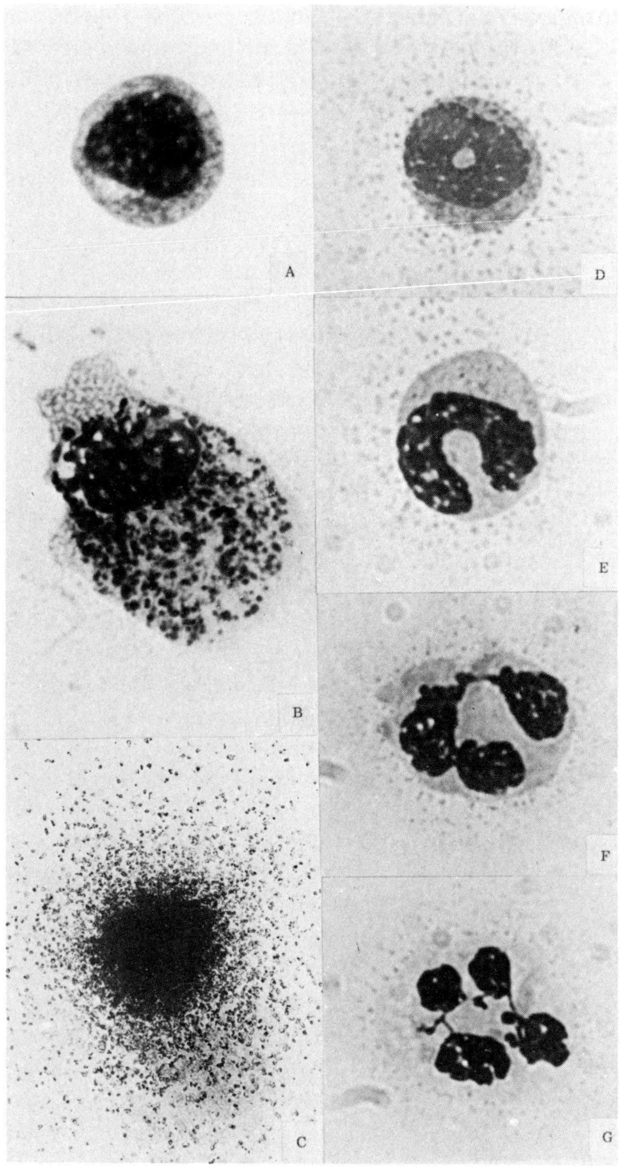

FIG. 2. Differentiation of myeloid leukaemic cells to non-malignant mature macrophages or granulocytes by normal myeloid differentiation-inducing protein. A, leukaemic cell; B, macrophage; C, colony of cells with macrophages; D–G, stages in differentiation to granulocytes (Fibach et al 1972).

**TABLE 2 Differentiation of normal and leukaemic myeloid cells after culturing with different normal myeloid regulatory proteins**

| | *Differentiation after culturing with:* | | | | |
|---|---|---|---|---|---|
| *Cell type* | *MGI-2 (IL-6)* | *IL-3* | *MGI-1GM (GM-CSF)* | *MGI-1G (G-CSF)* | *MGI-1M (M-CSF CSF-1)* |
| Normal precursor cells | + | + | + | + | + |
| Leukaemic clone II | + | − | − | ± | − |
| Leukaemic clone 7-M12 | − | + | + | − | − |
| Leukaemic WEHI-3B | − | − | − | + | − |

The study of different clones of myeloid leukaemic cells has also shown that in addition to $D^+$ clones there are differentiation-defective clones (called $D^-$ clones, Fibach et al 1973) (Fig. 3). Some $D^-$ clones are induced by a normal myeloid regulatory protein to an intermediate stage of differentiation, at which growth of the cells slows down; others could not be induced to differentiate even to this intermediate stage (see Sachs 1974, 1978, 1982, 1987c, Ichikawa et al 1976). Since normal differentiation inducer can induce differentiation to mature non-dividing cells in the $D^+$ clones, it has been suggested that $D^+$ clones are the early stages of leukaemia and that the formation of different types of $D^-$ clones may be later stages in the progression of malignancy (Sachs 1978). Studies on changes in the synthesis of cellular proteins in normal myeloid precursors and $D^+$ and $D^-$ myeloid leukaemic cells have shown that there have been changes from inducible to constitutive gene expression in the leukaemic cells, and that the most differentiation-defective $D^-$ clones showed the highest amount of constitutive gene expression (see Sachs 1980, 1987b). The protein changes during differentiation of normal myeloid precursors are induced as a series of parallel multiple pathways of gene expression. It can be assumed that normal differentiation requires synchronous initiation and progression of these pathways. The presence of constitutive instead of induced gene expression for some pathways can be expected to disrupt the coordination required for differentiation. Depending on the pathways involved, this asynchrony can then produce blocks in the induction and termination of the differentiation programme (Sachs 1980, 1987a,b).

## Different pathways for inducing differentiation

Studies with a variety of chemicals other than normal haemopoietic regulatory proteins have shown that many compounds can induce differentiation in

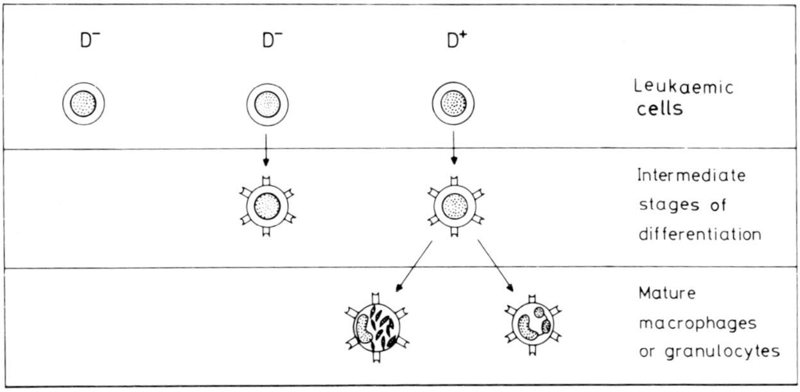


FIG. 3. Classification of different types of clones of myeloid leukaemic cells according to their ability to be induced to differentiate by normal myeloid differentiation-inducing protein. Some differentiation-defective ($D^-$) clones can be induced by normal differentiation-inducing protein to intermediate stages of differentiation, whereas other $D^-$ clones cannot.

$D^+$ clones of myeloid leukaemic cells. These include certain steroid hormones, chemicals such as cytosine arabinoside, adriamycin, methotrexate and other chemicals that are used today in cancer chemotherapy, and also X-irradiation. At high doses these compounds and X-irradiation kill cells, whereas at low doses they can induce differentiation. Not all these compounds are equally active on the same leukaemic clone (see Sachs 1978, 1982). A variety of chemicals can also induce differentiation in $D^-$ clones that are not induced to differentiate by a normal haemopoietic regulatory protein; in some clones induction of differentiation requires combined treatment with different compounds (see Sachs 1982). These results show that the $D^-$ clones have not lost all the genes for differentiation. In addition to certain steroids and chemicals used today in chemotherapy and irradiation, other compounds that can induce differentiation in myeloid leukaemic cells include insulin, retinoic acid, bacterial lipopolysaccharide, certain plant lectins and phorbol esters (see Sachs 1978, 1982, 1987b). The steroid hormones, insulin and retinoic acid are physiological compounds that can induce differentiation. It is probable that all myeloid leukaemic cells no longer susceptible to the normal regulatory proteins by themselves can be induced to differentiate by the appropriate combination of compounds.

Other types of leukaemic cells, such as erythroleukaemic cells, can also be induced to differentiate by various chemicals (Marks & Rifkind 1978). Erythropoietin, which induces the production of haemoglobin in normal erythrocytes, does not do so in these erythroleukaemias. These erythroleukaemias are thus like $D^-$ myeloid leukaemias that are not induced to differentiate by a normal myeloid regulatory protein. It has also been shown that some of the compounds that induce differentiation in leukaemic cells can induce differentiation in tumours derived from other types of cells (Gardner 1983).

Not only do different compounds induce differentiation but there are different ways of inducing differentiation. In myeloid leukaemic cells, some compounds induce differentiation by inducing the production of a normal differentiation-inducing protein in the $D^+$ leukaemic cells, whereas others, such as the steroid hormones, induce differentiation without inducing this protein (see Sachs 1987b). Different compounds can induce different parts of the differentiation programme. When combined treatment with different compounds is required, this produces by complementation the appropriate gene expression that is required for differentiation. This complementation of gene expression can occur at the level of either mRNA production or mRNA translation (see Sachs 1987b). Studies on different mutants of myeloid leukaemic cells have also shown that the constitutive protein changes which inhibit differentiation by the steroid hormone dexamethasone are different from the constitutive protein changes that inhibit differentiation by a normal differentiation-inducing protein (see Sachs 1987b). These experiments have identified different pathways of gene expression for inducing differentiation, and have also shown that genetic changes which inhibit differentiation by one compound need not affect differentiation by another compound that uses alternative pathways.

## Chromosome changes, cell differentiation and bypassing of genetic defects in the suppression of malignancy

Evidence has been obtained with various types of tumours, including sarcomas (Sachs 1974), teratocarcinomas (see Gardner 1983) and myeloid leukaemias (Sachs 1974, 1978), that malignant cells have not lost all the genes that control normal growth. This was first shown in sarcomas by the finding that it was possible to reverse the malignant to a non-malignant phenotype with a high frequency in cloned sarcoma cells whose malignancy had been induced by chemical carcinogens, X-irradiation, or by a tumour-inducing virus (Fig. 4) (Rabinowitz & Sachs 1968, 1970a,b). In sarcomas induced after transformation of normal fibroblasts in culture with chemical carcinogens (Berwald & Sachs 1963) or X-irradiation (Borek & Sachs 1966), this reversibility of malignancy included reversion to the limited lifespan of normal fibroblasts (Rabinowitz & Sachs 1970b). Chromosome studies on normal fibroblasts, sarcomas, revertants from sarcomas which had regained a non-malignant phenotype, and re-revertants, showed that the difference between these malignant and non-malignant cells is controlled by the balance between genes located on specific chromosomes that at that time were called genes for

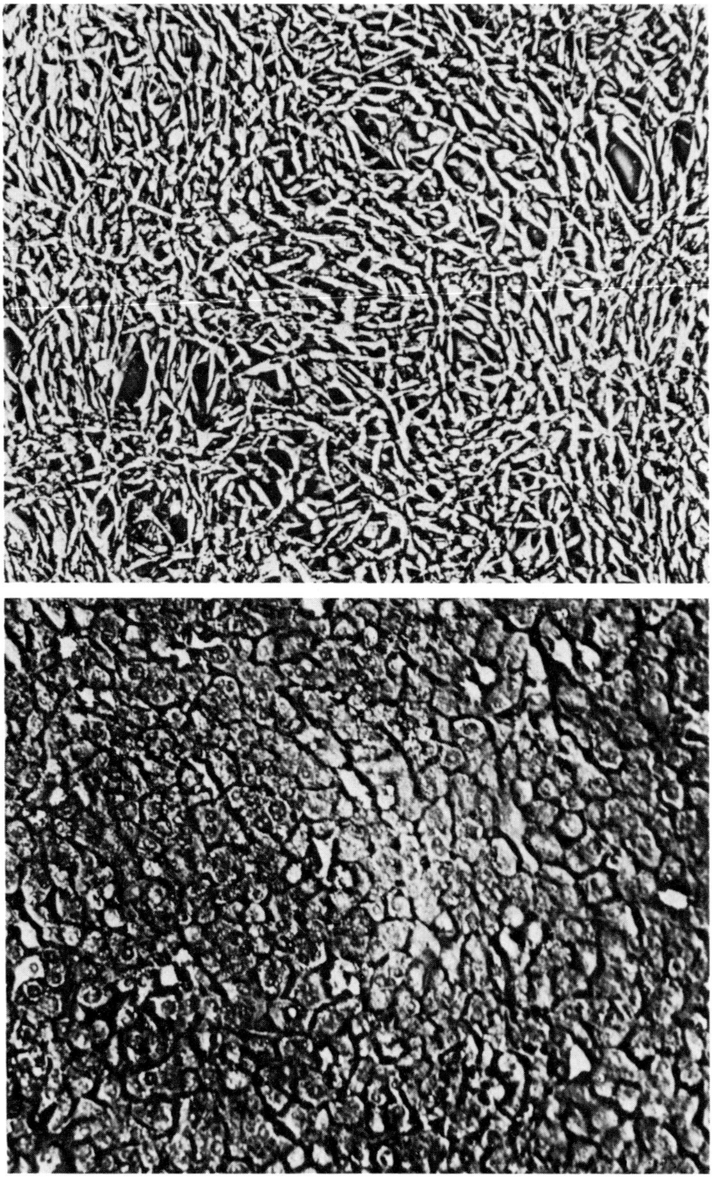

FIG. 4. Cultures of sarcoma cells (top) and a non-malignant revertant (bottom) (Rabinowitz & Sachs 1968).

expression (E) and genes for suppression (S) of malignancy (Fig. 5) (Rabinowitz & Sachs 1970a, Hitotsumachi et al 1971, Yamamoto et al 1973, Sachs 1974). When there is enough S to neutralize E, malignancy is suppres-

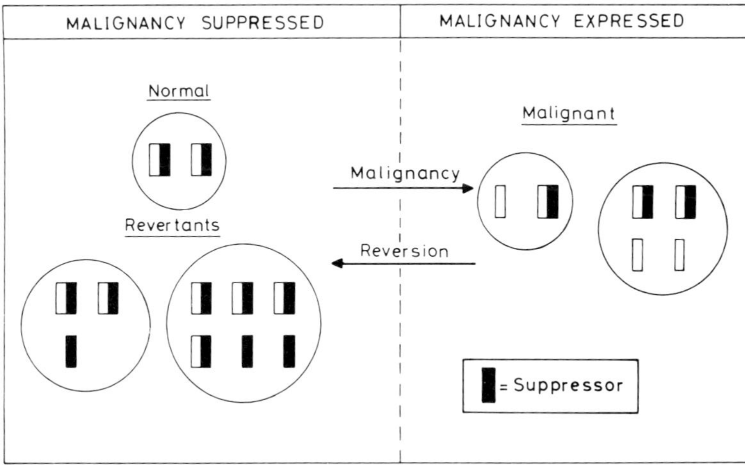


FIG. 5. The expression and suppression of malignancy by chromosome segregation resulting in a change in gene dosage owing to a change in the balance of specific chromosomes. From experiments with normal fibroblasts, sarcoma cells and their non-malignant revertants. Genes for expression □ and suppression ■ of malignancy (Rabinowitz & Sachs 1970a).

sed, and when the amount of S is not sufficient to neutralize E, malignancy is expressed. Thus, in addition to genes for expression of malignancy (E) (oncogenes), there are S genes that can suppress the action of oncogenes (see Sachs 1974, 1987a).

The suppression of malignancy in sarcomas (see Sachs 1974) was obtained by chromosome segregation, resulting in a change in gene dosage due to a change in the balance of specific chromosomes. This occurred without hybridization between different types of cells. The non-malignant cells were thus derived from the malignant ones by genetic segregation. Suppression of malignancy associated with chromosome changes including changes in gene balance has also been found after hybridization between different types of cells (see Klein 1981, Stanbridge 1984, Benedict et al 1984, Harris 1985). These studies on cell hybrids have led to similar conclusions to those obtained from the reversal of malignancy in sarcomas without hybridization between different cell types.

The $D^+$ myeloid leukaemic cells also have an abnormal chromosome composition (Azumi & Sachs 1977), but suppression of malignancy in these cells was not associated with chromosome changes: it was obtained by induction of the normal sequence of cell differentiation by a normal myeloid regulatory protein. In this suppression, the stopping of cell multiplication by inducing differentiation to mature cells bypasses genetic changes in the requirement for the normal growth inducer, and a block in the ability of growth inducer to

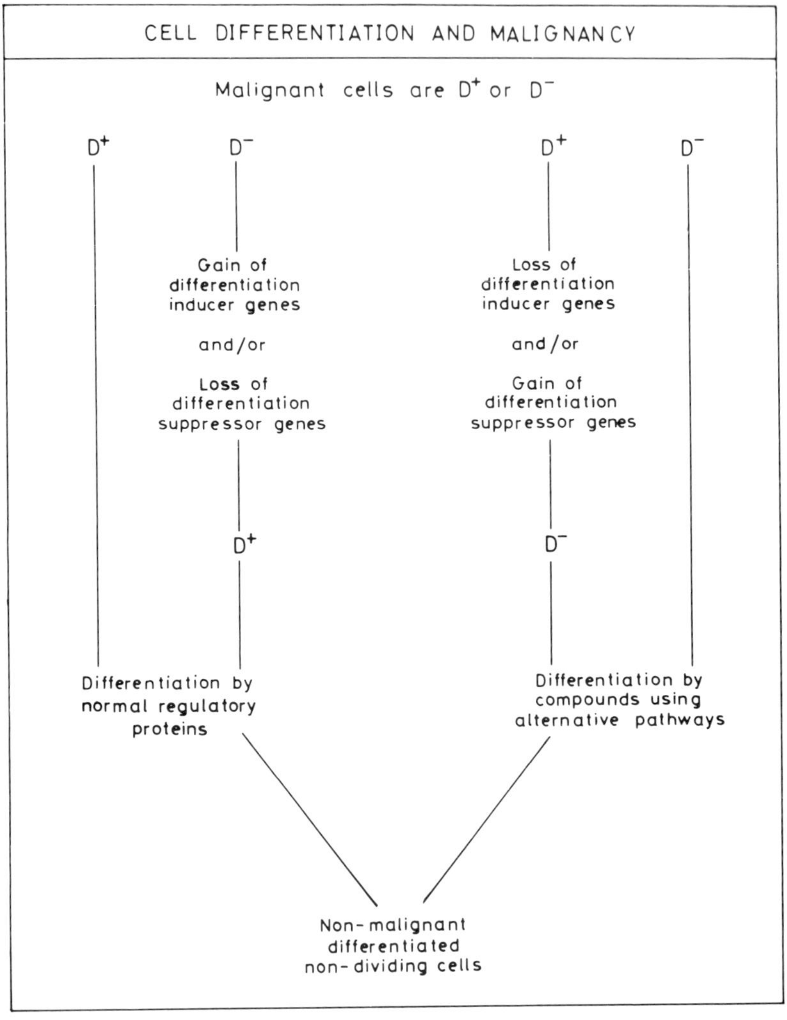


FIG. 6. Suppression of malignancy by inducing differentiation can be achieved in different ways. Malignant cells can ($D^+$) or cannot ($D^-$) be induced to differentiate to mature non-dividing cells by normal regulatory proteins. The $D^-$ cells can, however, be induced to differentiate by compounds that induce differentiation by alternative pathways. Chromosome changes that change the balance between genes for induction and genes for suppression of differentiation can change cells from $D^-$ to $D^+$ and *vice versa*.

induce differentiation inducer, that produced the malignant phenotype. Genetic changes which make cells defective in their ability to be induced to differentiate by the normal differentiation inducer occur in the evolution of myeloid leukaemia. But even these $D^-$ cells can be induced to differentiate by other compounds, either singly or in combination, that can induce the differentiation programme by alternative pathways. Also in these cases the stopping of cell multiplication by inducing differentiation by these alternative pathways bypasses the genetic changes that inhibit response to the normal differentiation inducer. This bypassing of genetic defects is presumably also the mechanism for the suppression of malignancy via the induction of differentiation in erythroleukaemias.

Studies on the chromosomes of myeloid leukaemic cells have also shown that the change from $D^-$ to $D^+$ and *vice versa*, i.e. the ability to be induced to differentiate to mature non-dividing cells by a normal myeloid regulatory protein, is controlled by the balance between genes that allow induction of differentiation and genes that suppress differentiation (Azumi & Sachs 1977). It was then shown in hybrids between different cell types that chromosome changes can suppress malignancy by restoring the ability of the cells to be induced to differentiate to non-dividing cells *in vivo* in a location in the body where the cells are exposed to what is presumably the normal differentiation inducer (Stanbridge 1984, Harris 1985). The appropriate chromosome changes also change these hybrid cells from $D^-$ to $D^+$. Chromosome changes can thus change malignant cells from $D^-$ to $D^+$, so that the cells can then be induced to differentiate when exposed to normal differentiation inducer.

It can, therefore, be concluded from studies on the molecular regulators of growth and differentiation in normal development, changes in the normal development programme in tumour cells, and suppression of malignancy that (1) malignancy can be suppressed by inducing differentiation either with or without genetic changes in the malignant cells (Fig. 6), (2) this suppression does not have to restore all the normal controls, and (3) genetic defects that give rise to malignancy can be bypassed and their effects nullified by inducing differentiation which stops cells from multiplying.

*Acknowledgements*

This research is supported by the Ebner Foundation for Leukemia Research, the National Foundation for Cancer Research, Bethesda, the Jerome A. and Estelle R. Newman Assistance Fund, and the Arly Winton Cancer Research Fund.

## References

Azumi J, Sachs L 1977 Chromosome mapping of the genes that control differentiation and malignancy in myeloid leukemic cells. Proc Natl Acad Sci USA 74:253–257

Benedict WF, Weismann BE, Mark C, Stanbridge EJ 1984 Tumorigenicity of human HT1080 fibrosarcoma × normal fibroblast hybrids: chromosome dosage dependency. Cancer Res 44:3471–3479

Berwald Y, Sachs L 1963 In vitro cell transformation with chemical carcinogens. Nature (Lond) 200:1182–1184

Blatt C, Goldberg R, Aberdam D, Sachs L 1987 Homeobox gene rearrangement in myeloid leukemic cells. Third Annual Meeting Oncogenes, Fredrick, Md, p 364

Blatt C, Aberdam D, Schwartz R, Sachs L 1988 DNA rearrangement of a homeobox gene in myeloid leukemic cells. EMBO (Eur Mol Biol Organ) J 7:4283–4290

Blatt C, Sachs L 1988 Deletion of a homeobox gene in myeloid leukemias with a deletion in chromosome 2. Biochem Biophys Res Commun 156:1265–1270

Borek C, Sachs L 1966 In vitro cell transformation by x-irradiation. Nature (Lond) 210:276–278

Bradley TR, Metcalf D 1966 The growth of mouse bone marrow cells in vitro. Aust J Exp Biol Med Sci 44:287–300

Clark SC, Kamen R 1987 The human hematopoietic colony-stimulating factors. Science (Wash DC) 236:1129–1237

Fibach E, Landau T, Sachs L 1972 Normal differentiation of myeloid leukemic cells induced by a differentiation-inducing protein. Nat New Biol 237:276–278

Fibach E, Hayashi M, Sachs L 1973 Control of normal differentiation of myeloid leukemic cells to macrophages and granulocytes. Proc Natl Acad Sci USA 70:343–346

Gardner RL 1983 Teratomas in perspective. Cancer Surv 2:1–19

Ginsburg H, Sachs L 1963 Formation of pure suspension of mast cells in tissue culture by differentiation of lymphoid cells from the mouse thymus. J Natl Cancer Inst 31:1–40

Gootwine E, Webb CG, Sachs L 1982 Participation of myeloid leukemic cells injected into embryos in hematopoietic differentiation in adult mice. Nature (Lond) 299:63–65

Harris H 1985 The suppression of malignancy in hybrid cells: The mechanism. J Cell Sci 79:83–94

Hitotsumachi S, Rabinowitz Z, Sachs L 1971 Chromosomal control of reversion in transformed cells. Nature (Lond) 231:511–514

Ichikawa Y, Pluznik DH, Sachs L 1966 In vitro control of the development of macrophage and granulocytes colonies. Proc Natl Acad Sci USA 56:488–495

Ichikawa Y, Maeda N, Horiuchi M 1976 In vitro differentiation of Rauscher virus induced myeloid leukemic cells. Int J Cancer 17:789–796

Klein G 1981 The role of gene dosage and genetic transposition in carcinogenesis. Nature (Lond) 194:313–318

Lotem J, Sachs L 1981 In vivo inhibition of the development of myeloid leukemia by injection of macrophage and granulocyte inducing protein. Int J Cancer 28:375–386

Lotem J, Sachs L 1986 Regulation of cell surface receptors for different hematopoietic growth factors on myeloid leukemic cells. EMBO (Eur Mol Biol Organ) J 5:2163–2170

Lotem J, Sachs L 1987 Regulation of cell surface receptors for hematopoietic differentiation-inducing protein MGI-2 on normal and leukemic myeloid cells. Int J Cancer 40:532–539

Lotem J, Sachs L 1988 In vivo control of differentiation of myeloid leukemic cells by recombinant granulocyte-macrophage colony stimulating factor and interleukin 3. Blood 71:375–382

Marks P, Rifkind RA 1978 Erythroleukemia differentiation. Annu Rev Biochem 47:419–448

Metcalf D 1985 The granulocyte-macrophage colony-stimulating factors. Science (Wash DC) 299:16–22

Olsson I, Sarngadharan MG, Breitman TR, Gallo RC 1984 Isolation and characterization of a T lymphocyte-derived differentiation inducing factor for the myeloid leukemic cell line HL-60. Blood 63:510–517

Pluznik DH, Sachs L 1965 The cloning of normal 'mast' cells in tissue culture. J Cell Comp Physiol 66:319–324
Pluznik DH, Sachs L 1966 The induction of clones of normal 'mast' cells by a substance from conditioned medium. Exp Cell Res 43:553–563
Rabinowitz Z, Sachs L 1968 Reversion of properties in cells transformed by polyoma virus. Nature (Lond) 220:1203–1206
Rabinowitz Z, Sachs L 1970a Control of the reversion of properties in transformed cells. Nature (Lond) 225:136–139
Rabinowitz Z, Sachs L 1970b The formation of variants with a reversion of properties of transformed cells. V. Reversion to a limited life span. Int J Cancer 6:388–398
Sachs L 1974 Regulation of membrane changes, differentiation and malignancy in carcinogenesis. The Harvey Lectures, Academic Press, New York, vol 68:1–35
Sachs L 1978 Control of normal cell differentiation and the phenotypic reversion of malignancy in myeloid leukaemia. Nature (Lond) 274:535–539
Sachs L 1980 Constitutive uncoupling of pathways of gene expression that control growth and differentiation in myeloid leukemia: a model for the origin and progression of malignancy. Proc Natl Acad Sci USA 77:6153–6156
Sachs L 1982 Normal developmental programmes in myeloid leukaemia: regulatory proteins in the control of growth and differentiation. Cancer Surv 1:321–342
Sachs L 1987a Cell differentiation and by-passing of genetic defects in the suppression of malignancy. Cancer Res 47:1981–1986
Sachs L 1987b The molecular regulators of normal and leukaemic blood cells. The Wellcome Foundation Lecture 1986. Proc Soc Lond B Biol Sci 231:289–312
Sachs L 1987c The molecular control of blood cell development. Science (Wash DC) 238:1374–1379
Shabo Y, Sachs L 1988 Inhibition of differentiation and affinity purification with a monoclonal antibody to a myeloid cell differentiation-inducing protein. Blood 72:1543–1549
Shabo Y, Lotem J, Rubinstein M et al 1988 The myeloid blood cell differentiation-inducing protein MGI-2A is interleukin 6. Blood 72:2070–2073
Stanbridge EJ 1984 Genetics analysis of tumorogenicity in human cell hybrids. Cancer Surv 3:335–350
Tomida M, Yamamoto-Yamaguchi Y, Hozumi M 1984 Purification of a factor inducing differentiation of mouse myeloid leukemic M1 cells from conditioned medium from mouse fibroblast L929 cells. J Biol Chem 259:10978–10982
Yamamoto T, Rabinowitz Z, Sachs L 1973 Identification of the chromosomes that control malignancy. Nat New Biol 243:247–250

## DISCUSSION

*Harris:* In principle, there should be situations where you do actually correct the genetic defect. If there are not too many defects, experiments designed to correct them should throw some light on what causes malignancy. If you transfect pieces of DNA, you narrow the range of your observations and are able to look at one particular parameter at a time. I agree that you don't need to do any of this in order to turn a malignant cell into a non-malignant one.

*Sachs:* In studying the suppression of malignancy, one question is, to what extent can we suppress malignancy in the sense of reversing the genetic abnormalities that made the cells malignant? Another question is, to what extent can we suppress malignancy by switching on programmes such as differentiation programmes which change the malignant cells into non-dividing mature cells without correcting the original genetic defects? One of the points I emphasized was that there are ways to suppress malignancy which bypass the genetic abnormalities that caused the malignancy.

*Adams:* I was pleased to hear that you have cloned a new type of differentiation factor. Is it a novel lymphokine or one that is already known in another system?

*Sachs:* This is the myeloid cell differentiation factor that we originally described about 10 years ago, which we have called MGI-2.

*Sporn:* Has the sequence of that been published?

*Sachs:* Yes, in Shabo et al (1988).

*Adams:* So the factor works on normal myeloid precursors?

*Sachs:* Yes.

*Varmus:* Is the product of the cloned gene made in bacteria and does it induce the same events as does the purified protein?

*Sachs:* We have characterized the protein in collaboration with the Genetics Institute in Boston. The recombinant protein made by either bacteria or mammalian cells has the same effect as the purified protein.

*Stanbridge:* How do experimentally induced or naturally occurring leukaemias in mice behave under these differentiation-inducing conditions? Can you induce differentiation of naturally occurring murine leukaemias?

*Sachs:* I mentioned that differentiation of the leukaemic cells and the accompanying suppression of malignancy can be induced in mice *in vivo*. The myeloid leukaemic cells that respond *in vitro* also respond *in vivo*: this includes leukaemic cells from a 'naturally' occurring myeloid leukaemia.

*Adams:* What do you make of the great difficulty in establishing most human leukaemias as cell lines *in vitro*? It has been easy to say that we do not know the right culture conditions, but as more and more haemopoietic factors become available, that seems a less likely explanation.

*Sachs:* I suppose it depends under what conditions the cells grow *in vivo*. It could be something like a particular amino acid or other nutrient that nobody has tried. In addition, some cells need a matrix to grow, others do not. The localized conditions required by the leukaemic cells may not necessarily all be satisfied by taking the factors which are known at present and adding them to the culture dish. Also, if one wants to keep the cells growing, one has to ensure that they do not differentiate into non-dividing cells. I think it is a reasonable principle that if you really want to grow the cells, you are going to find a way of doing so.

*Noble:* In our studies on oncogene cooperativity in Schwann cells, if you take SV40 large T antigen and put it into a Schwann cell, you lose some growth factor requirements, as you would expect, but the Schwann cells are already immortal (Ridley et al 1988). If you put in both large T and *ras* protein, you get classical oncogene cooperativity. The cells become completely growth factor independent and their cell cycle time falls from 72 hours to 12 hours. If you use a temperature-sensitive T antigen, and turn it off by growing cells at 39°C, or if you inject *ras* protein into a normal cell, the effect of *ras* on the Schwann cell is to induce proliferation arrest. Thus, the nuclear oncogene changes the effect of the signal from *ras* within the cell. The question then, with regard to the effects of insulin and the steroid hormone, prednisolone, in leukaemic cells, is does this reflect an immortalization event within a cell giving aberrant responses to compounds that in a normal cell have nothing to do with those pathways of differentiation?

*Sachs:* In the case of prednisolone we have checked this in experiments to determine the effects of the steroid on the developmental programme of normal myeloid haemopoietic cells. You cannot add only the steroid, because in the absence of an appropriate haemopoietic regulatory protein the normal precursor cells die, so you have to add the steroid plus one of these proteins. In normal myeloid cell development the steroid modulates the pathway of differentiation that the cells follow, so there is an increased frequency of differentiation to granulocytes.

## References

Ridley AJ, Paterson HF, Noble M, Land H 1988 *ras*-mediated cell cycle arrest is altered by nuclear oncogenes to induce Schwann cell transformation. EMBO (Eur Mol Biol Organ) J 7:1635–1645

Shabo Y, Lotem J, Rubinstein M et al 1988 The myeloid blood cell differentiation-inducing protein MGI-2A is interleukin 6. Blood 72:2070–2073

# General discussion VI

## Critical events in tumour suppression

*Knudson:* It has always struck me as rather curious that formation of these cell hybrids really does suppress tumorigenicity, when we know there is a sequence of changes involved in the generation of a tumour. Some of us thought that even if you corrected the original defect that caused the tumour, all these other changes would continue to make it malignant. Can any of you who are working on suppression comment on whether this speaks for some single gene having primary importance over all others, which are really secondary changes and not critical, or whether it means that more than one event is necessary, and all these events must occur.

*Stanbridge:* I personally see no *a priori* reason why all chromosomal changes, for example, should necessarily contribute to the malignant status of a cell—there can be many chromosomal changes that have absolutely nothing whatsoever to do with malignancy. There are many examples in the experimental literature of aneuploid cells which are immortal but non-tumorigenic. For example, chromosomal changes following SV40 infection of human cells are quite dramatic, yet the vast majority of those cells go on to senesce and only a small proportion of those come out of the senescence crisis to form immortal cell lines—which are still non-tumorigenic in the vast majority of cases.

*Knudson:* It is an important question: are cancers single gene disorders, errors of metabolism, and can we think about treating them as such?

*Bodmer:* My answer is that the nude mouse model is a very inadequate model: you never get proper metastasis with human tumours in nude mice, so there are other factors involved. I don't think that by inoculation of a tumour into a mouse you are looking at anything that is analogous to the original tumorigenic process.

*Stanbridge:* That may be correct but in the vast majority of experimental systems that have been discussed at this meeting, we have been talking about suppression of tumorigenicity rather than suppression of malignancy. The latter is a much more difficult phenotype to assess.

If we are talking about somatic cell hybrids, we are talking about suppression of tumorigenicity. When one examines somatic cell hybrids there are situations, for example, with the pseudodiploid HT1080 human fibrosarcoma cell line, where if one fuses it with a normal fibroblast, in the vast majority of cases the hybrid behaves like a normal cell in that it has a limited proliferative lifespan. Segregants, if one likes to use that terminology, from that population arise that have gained the capacity for indefinite proliferation. These cells are non-tumorigenic. If one now grows them in culture for long periods of time, there is another segregation event, resulting in a tumorigenic segregant. Thus, there are different events for different phenotypes, but it is presumably a finite number, otherwise we would be in trouble.

*Bodmer:* We are talking about the total malignant process and the number of genes involved. Whether this is done in syngeneic models or however you do it, putting cells into a mouse is missing very significant parts of the full tumorigenic process. Secondly, if you make interspecies hybrids, unless in each case you do a careful analysis of single chromosome loss, you may be complementing several functions at once. Neither of those arguments allows you to say that you are complementing only one gene function and that that is the critical function in the malignant process or even for the tumorigenic process as studied in mouse inoculations.

*Harris:* I believe that there are at least two mechanisms by which malignancy is suppressed in hybrid cells. As far as I can see, all the tumours examined to date are caused by some genetic loss of function. Where malignancy is suppressed in hybrids, you may actually be complementing the specific genetic loss that is critical to determining the malignant phenotype. I think this may be what is happening when you introduce a single chromosome or a single gene into the cell.

The other form of suppression occurs when you fuse a malignant cell with a normal one, and the hybrid cell executes and brings to completion the terminal differentiation programme of the normal cell. The imposition of that terminal differentiation programme will override whatever panoply of genetic events might have been responsible for the generation of the malignant phenotype in the first place. That ought not to be too surprising, for two reasons. First, if tumour cells didn't throw off non-malignant differentiated variants, we would have no tumour pathology. If you look at fibrosarcomas, for example, they come in various grades: some fibrosarcomas make a lot of collagen, some don't, and there are mixed tumours. If you look for mitoses in fibrosarcomas, you will find very few mitoses in the regions where collagen is deposited; they are virtually all found in areas where there is little or no collagen. Those tumours that make little collagen grow more rapidly than the collagen-producing ones.

More powerful evidence comes from the classical experiments on crown gall tumours by Armin Braun (Braun & Wood 1976, Turgeon et al 1976). The normal cells of a tobacco plant have certain requirements for plant growth hormones when grown *in vitro*. You produce a crown gall tumour in a tobacco plant and take out the crown gall tumour cells. You can regenerate a whole plant from a single crown gall tumour cell and grow an apparently perfectly normal plant from it. Initially it was argued that therefore malignancy couldn't be determined by classical mutations. But when cells from the apparently normal plant grown from a single crown gall tumour cell were explanted, the cells remembered that they were tumour cells and grew *in vitro* without added plant growth hormones. Every aspect of the malignant phenotype of the crown gall tumour cell was overridden by the process of differentiation.

*Stanbridge:* This is a very important point. When one is dealing with somatic cell hybrids and whole cell fusions one is unable to dissociate the non-tumorigenic phenotype from an induction of the programme of differentiation. Interestingly, the programme that is induced always seems to be that for the phenotype of the normal cell, irrespective of the origin of the tumorigenic cell (Stanbridge et al 1982). The tumorigenic segregants of those hybrids, both in culture and in the athymic nude mouse, are relatively undifferentiated. The difference, in cytogenetic terms, between the tumorigenic segregant and its non-tumorigenic progenitor may be as discrete as the loss of a single normal chromosome. If one introduces that proposed single chromosome into the tumorigenic parental cell, tumorigenicity is suppressed. But, with the limited experience that we have, we don't see obvious signs of differentiation. Thus, on the one hand we have suppression and differentiation in whole cell hybrids, and on the other hand we have possible suppression without obvious signs of differentiation in microcell hybrids. It is possible that the differentiation is somewhat of a red herring in that the non-tumorigenic whole cell hybrid stops dividing as a consequence of the tumour suppressor effect, and somehow the differentiation programme, the genetic information for which one has also introduced with all the other normal chromosomes, is also switched on. If the hybrid doesn't stop dividing in the intact animal, for whatever reason, that differentiation programme does not come into effect.

*Stanbridge:* We looked for the other markers of differentiation, too.

*Harris:* There are very few tumour cells whose pattern of terminal differentiation has been sufficiently well worked out in biochemical terms for one to know whether one has complemented a block to differentiation. There are a few tumours, for example melanomas, where we know something about the biochemistry of terminal differentiation, but not many.

*Stanbridge:* I agree with that but I do believe it's still an open question.

*Klein:* Some key words in this discussion are 'block to differentiation' in malignant cells and 'overriding by differentiation' in suppression. A third key point is that activated oncogenes only transform during very specific differentiation windows. If you shift the cell out of that window, that cell can terminally differentiate. The best examples are the temperature- sensitive *src* and *erbB* experiments of Holtzer, Bottinger and others, where differentiation is blocked at the permissive temperature, but down regulation of the oncogene for a little while permits the cell to bypass the block and differentiate. There is probably only one rate-limiting change, once you overcome that key block, it doesn't matter how many secondary changes there are. The cell line HL60 has an amplified *myc* gene and many chromosomal changes. If you add a non-specific agent, such as DMSO, the amplified *myc* gene is switched off and there is complete differentiation. In the microcell transfer experiment, you might be introducing a gene that shifts the phenotype slightly to a point where it will be responsive to whatever control prevails. The kind of switch I described, where a Burkitt's lymphoma cell has switched to an immunoblastic cell and is no longer malignant was another example of a phenotypic change that has led to responsiveness to the prevailing growth control signals. We should be cautious about genetic complementation, since we are not dealing with naked genomes as in the prokaryotes.

*Sachs:* This may be, again, a question of terminology. When a malignant cell is changed back into a non-malignant cell, this does not necessarily mean that the genetic defects which made the cell malignant to begin with are corrected. If you can switch on the differentiation programme, even without cell fusion, these cells are still chromosomally abnormal. There are a number of steps which make the cell malignant, but if you stop the cells dividing by inducing differentiation, you will bypass the genetic abnormalities which made the cells malignant.

There is no conflict in saying that a number of things have to happen when cells change from benign to malignant to metastatic, and that you can stop the cells forming a tumour by stopping the cells dividing. In the myeloid leukaemia cells, we can induce differentiation to mature non-dividing cells. We can also induce differentiation to a stage when the cells require an external source of growth factor for cell viability. If the cells are then in a part of the body or under conditions where the growth factor is not present, the cells are no longer malignant because they die. The growth factor provides cell viability without in any way correcting the genetic abnormalities which made the cells malignant.

*Marks:* In both murine erythroleukaemia cells and HL60 cells, you can bypass the block to differentiation and suppress tumorigenicity by exposing cells to HMBA or related polar compounds. These cells can also be induced to express differentiated genes without suppressing the capacity of cells to divide. I think we should assay for cell proliferation, not only expression of genes characteristic for a differentiated phenotype.

*Weinberg:* All these replies to Al's question have not addressed the issue that he raised. I think the question he asked was the following. If you grant that there are a series of successive genetic changes that create the final tumour cell phenotype, one occurring after another, and that the tumorigenic phenotype requires the collaboration of all these different changes, (given the limitations of the mouse tumorigenicity test), by reversing genetically one of those early changes will you cause the tumour cell to lose its tumorigenicity? Or will it still be tumorigenic because that early genetic change is not required for the maintenance of tumorigenicity? Has the tumour cell developed other mechanisms to maintain tumorigenicity which no longer require the initiating step?

*Wyke:* The cell hybrid experiments cannot answer that question.

*Weinberg:* The cell hybrid studies address the issue of whether a genetic lesion continues to be required for the maintenance of tumorigenic phenotype. They don't address whether that lesion was an early or late one. (A late lesion is just as aetiological as is an early one.)

To my knowledge there are not yet any well developed examples where an initiating lesion is no longer required for maintenance of the tumorigenic phenotype, but in theory it is perfectly possible. There are examples where tumours which seem to have been initiated by a *ras* oncogene point mutation, later in tumour progression develop a second activated *ras* oncogene. In that case, you could imagine that the initiating change is no longer required for maintenance of the phenotype, since a second, physiologically equivalent, genetic event has occurred.

*Wyke:* I would like to know how viruses, such as herpes virus, cause tumours. If we knew this, we might be able to address the question of whether initiating lesions are later dispensable. It is conceivable that in some cases of viral neoplasia there was an initiating event that is no longer required.

*Weinberg:* A so-called hit-and-run event.

*Lau:* Another example is the plasmid-induced hit-and-run tumorigenesis in Chinese hamster embryo fibroblast cells (CHEF) (Lau et al 1985); others are transformations following infection with herpes simplex virus (Galloway & McDougall 1983) and with adenovirus type 12 (Kuhlmann et al 1982).

With the CHEF cells, transfection with pSV*gpt* or pSV*neo* causes extensive chromosomal rearrangements. These cells subsequently become tumorigenic, even in the total absence of any foreign DNA. In transformation by herpes simplex virus, no unique set of viral genes seems to be consistently retained or expressed in the transformed cells, and similar results have been obtained with human cervical carcinoma, suggesting that viral DNA is not needed for maintenance of the transformed phenotype. Tumorigenicity can also be maintained after the loss of all viral DNA from tumour cells transformed by adenovirus. Furthermore, there is no conclusive evidence that initiation of neoplasia is mediated by a viral protein (Galloway et al 1984).

*Knudson:* Suppose a Wilms' tumour shows N-*myc* amplification and *ras* mutations, but hybridization with chromosome 11 suppresses tumorigenicity. Is that change so critical that all others don't matter? Or does it mean that all of the mutations are needed for tumorigenicity and correcting only one of them is enough? If correcting the initial lesion or correcting one of several is adequate, then there is some hope of reversing cancer by correcting one gene instead of looking at the whole tumour. By tumour suppressor genes do we mean genes that have that potential?

*Klinger:* We do not know the answer, but my guess is that 'correcting' one of the lesions, not necessarily the initial one, would reverse the malignant process. It is clear from the cell hybrid studies we have been discussing that it is not necessary to correct all that is wrong with a malignant cell to stop it from being tumorigenic. Non-tumorigenic hybrids are by no means normal, since they express several abnormal phenotypes but still fail to produce tumours. I find this very encouraging, because the independent segregation of tumorigenicity from these abnormal phenotypes, which were formerly believed to be closely related to tumorigenicity, means that we do not have to worry about them, we have to be concerned only with those that are directly related to making the cell tumorigenic.

The evidence collecting in favour of Al Knudson's hypothesis that several changes are critical, means that in cancer therapy we should concentrate on identifying those changes that are critical for the abnormal pathways of a malignant cell, since altering only one of these aberrant pathways is likely to be enough to reverse the malignant process. Once we understand these pathways, it is likely that methods could be devised for either blocking or correcting them. The results from the somatic cell hybrid and chromosome transfer studies suggest that the product of a single or very limited number of normal genes is enough to suppress the tumorigenic phenotype. This is supported by many other recent developments, some of which we have heard of here, for example, the existence of 'suppressor' genes and the insight being gained into the function of oncogenes. I suspect that in some cases the same gene, when expressed inappropriately, will be both an oncogene and a suppressor, depending on the tissue or developmental stage at which this happens, analogous to what appears to be happening in the TGF-β system. This could greatly simplify the development of therapeutic approaches.

If this separation of the important from less important genetic mechanisms is to be realized, we must remember that we are dealing with a wide range of tumorigenic phenotypes and must consequently present our findings in as quantitative a manner as possible. For example, the data on the tumorigenic characteristics of parental cells and of hybrids is inadequate in some reports. The significance of results from some molecular studies of tumours, such as reports on the loss of heterozygosity, are difficult to determine without cytogenetic or independent molecular determinations of the chromosome complement of the cells studied, or without an adequate description of the histopathological and related findings.

*Adams:* There seems to be a degree of consensus that most studies with tumour cell–normal cell hybrids tell one very little about the nature of the tumorigenic events. It seems that the suppression of tumorigenesis in such hybrids may say almost nothing about whether a recessive oncogene was the cause of the tumorigenesis in the tumour cell partner of the hybrid.

*Harris:* It really depends on what you think the homozygous recessive genetic lesions are doing. I believe that the natural kinetic steady-state of all cells, eukaryotic and prokaryotic, is exponential multiplication, because that's what evolution selects for. What brings exponential multiplication under control in the developing embryo is some form of differentiation.

My personal guess is that these homozygous recessives that Web Cavenee and other people are talking about result in blocks to differentiation. I don't, of course, mean the complete absence of differentiation, but a block at some critical point where a mechanism comes into play (a brake) that normally takes cells out of their natural kinetic steady state (exponential multiplication) and thus allows them to undergo further differentiation. I believe the progressive multiplication of malignant cells is a secondary consequence of a block to differentiation.

*Adams:* While we all accept that there are genes which act as recessive oncogenes, the question is whether firm conclusions about recessive oncogenes can be drawn from the fusion of tumour cells and normal cells. If one takes a tumour cell about which nothing is known, fuses it to a normal cell and finds that the hybrid is not tumorigenic, can one conclude from that information whether a recessive oncogene was involved?

*Harris:* You have to do what is technically possible. When I did the first experiments of this kind with Jack Miller, we didn't know what to expect. I didn't believe that cancer could be caused by dominant mutations, and I still don't believe it. But we obtained an interesting result and the experiment has had some interesting consequences. The fact that malignancy is suppressed doesn't necessarily tell you anything about the initial lesions that determined the malignancy. But if you explore how the suppression might be brought about, it can be enlightening.

*Stanbridge:* I think the point is being lost here. It has been stated that in the somatic cell hybrid experiments there is an initial suppression followed by re-expression of tumorigenicity, presumably as a consequence of discrete loss of genetic information. There is still a significant amount of the genetic information of the normal parent cell present in the tumorigenic hybrid and it is every bit as tumorigenic as the original tumorigenic parental cell. So, I believe it is a situation of discrete events. It should not be looked upon as something totally random and non-specific. It is a case of specific chromosome loss associated with re-expression of tumorigenicity. That's why the somatic cell hybrids have been particularly useful, because they point one in the direction of the chromosome(s) worth looking at.

*Nossal:* Carrying that line of argument one step further, is there anything we can do to take advantage of the experiments of Nature? We have heard evidence that specific chromosome losses lead to certain types of tumours. Can we map that information on to the large body of information that exists about growth factors and differentiation factors? Is there anything that the commonality can tell us about the grand pattern of the motifs of factor actions on the different cells and tissues of the body?

*Sachs:* That is a very good point. I fully agree with Henry about the importance of differentiation: the problem of cancer is a problem in developmental biology. Normal development depends on a variety of molecular regulators, growth factors and differentiation factors, and the interactions between them. If there are too many dividing cells and not enough differentiated non-dividing cells, this uncouples the normal relationship between growth and differentiation and can result in malignancy. Therefore, by identifying the growth and differentiation factors, and finding out which genetic blocks in development lead to malignancy, one can get at the problem of tumorigenesis and find out what these genes actually do.

*Adams:* But some profound effects on cell proliferation may simply reflect slight changes in the balance of transcription factors. After all, the normal cell used in cell fusions may be at a different stage of differentiation, or even from a different lineage. Perhaps something as subtle as a 10% change in the proportion of cell divisions that lead to differentiation versus cell renewal might convert a cell from tumorigenic to normal.

*Harris:* The real puzzle is not why cells multiply, but why cells do not multiply, what restrains cell multiplication. If that doesn't get you into differentiation, I don't know what will.

## References

Braun AC, Wood HN 1976 Suppression of the neoplastic state with the acquisition of specialized functions in cells, tissues, and organs of crown gall teratomas of tobacco. Proc Natl Acad Sci USA 73:496–500

Galloway DA, McDougall JK 1983 The oncogenic potential of herpes simplex viruses: evidence for a 'hit-and-run' mechanism. Nature (Lond) 302:21–24

Galloway DA, Nelson JA, McDougall JK 1984 Small fragments of herpes virus DNA with transforming activity contain insertion sequence-like structures. Proc Natl Acad Sci USA 81:4736–4740

Kuhlmann I, Achten S, Rudolph R, Doerfler W 1982 Tumor induction by human adenovirus type 12 in hamsters: loss of the viral genome from adenovirus type 12-induced tumor cells is compatible with tumor formation. EMBO (Eur Mol Biol Organ) J 1:79–86

Lau CC, Gadi IK, Kalvonjian S, Anisowicz A, Sager R 1985 Plasmid-induced "hit-and-run" tumorigenesis in Chinese hamster embryo fibroblast (CHEF) cells. Proc Natl Acad Sci USA 82:2839–2843

Stanbridge EJ, Channing JD, Doersen C-J et al 1982 Human cell hybrids: analysis of transformation and tumorigenicity. Science (Wash DC) 215:252–259

Turgeon R, Wood HN, Braun AC 1976 Studies on the recovery of crown gall tumor cells. Proc Natl Acad Sci USA 73:3562–3564

# Final discussion

## General models for the generation and suppression of malignancy

*Bodmer:* We have developed a model to integrate what is known about genetic dominance and recessiveness, positive and negative regulation and the different mechanisms involved in the development of malignancy (Fig. 1). Malignancy involves a number of steps which can occur in a variety of sequences. So there are genes affecting growth factors, their receptors, differentiation factors and their receptors, positive and negative regulators and steps in the signalling process. We are talking about the various ways in which those different genes can act in the malignant process. The dominantly acting oncogenes have been largely associated with growth factors and their receptors; switching them on where they are not normally expressed. This can be linked to positive regulation at the DNA level.

A simple assumption is that the normal state is to be differentiated and not growing. If you switch on growth when it is not usually switched on, you have a step towards malignancy. If you switch off differentiation when it should be occurring, you have a step towards malignancy. You can switch on growth by positive activation and you can switch off growth by suppression. The simplest interpretation is that when you switch something on, you only need to change one thing and that should be a dominant genetic effect. If you have a negative control or suppressor, to switch that off you need a recessive effect. If you want to switch off differentiation due to a receptor, for instance, you have to knock out both gene activities—so it is recessive.

If you have positive activation, you also have recessiveness and if you have suppression, dominance. You can block differentiation by switching on only one copy of a gene that is a negative regulator; switching off differentiation in that way will be genetically dominant. If, however, activation is required to switch off differentiation, the process will be recessive. So you can draw a complete table of these major functions in terms of whether something is dominant or recessive and whether you are looking at activation or suppression (Table 1, Bodmer 1988).

*Sachs:* One point which we have not discussed is the cause of genetic instability in tumours.

*Harris:* A more basic issue is, what is the cause of aneuploidy? Why do the chromosomes become scrambled in some situations and not in others? Scrambling of the chromosomes is a very important experimental consideration. If you interpolate a gene or virus and produce some heritable effect in the cell, it

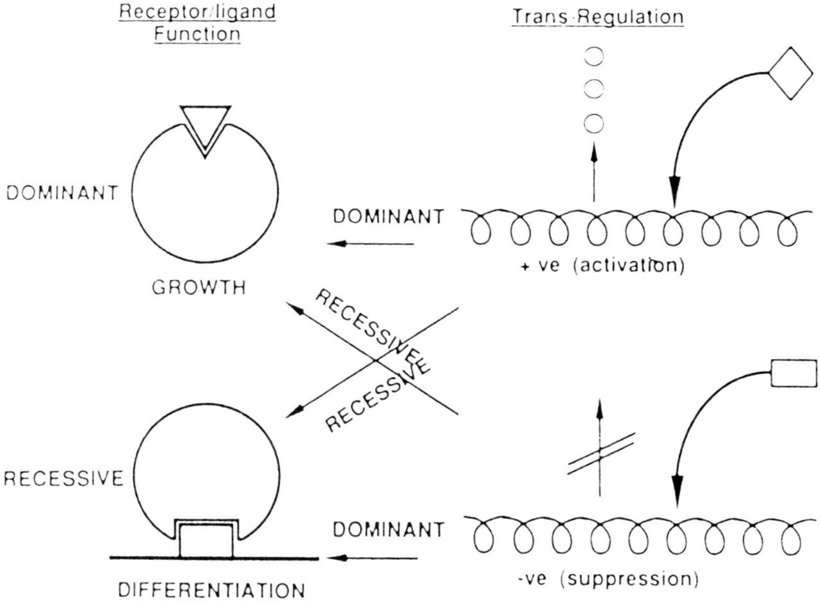


FIG. 1. (*Bodmer*) Scheme for dominant and recessive changes in cancer progression based on receptor–ligand function and trans-regulating factors.

behoves you to look at the chromosomes. If, after transfection, the cell has something like the karyotype you started with, you might be in a position to talk about what the interpolated gene is doing. But if you have a totally altered karyotype, then you are dealing with a genetically different cell, and any conclusions about what the interpolated gene might be doing in determining malignancy become extremely precarious.

*Adams:* Do we have to regard every karyotypic change found in a cancer cell as having some selective advantage?

*Knudson:* I'd say no. I think most cytogeneticists would say there are some changes that are probably aetiological, some involved in metastasis and progression, and some just accidents resulting from the ability of tumour cells to make errors.

*Bodmer:* Doesn't that depend on the relative frequency of events? If you have a relatively unstable karyotype, then the rate at which variants are produced is so high that some of those variations are going to be chance variations that have nothing to do with the selective process. If you have a more stable karyotype, a persistent change against that background is unlikely not to be a result of selection.

*Adams:* That is what I intuitively feel, but is there some mathematical rule with which to evaluate this? As an example, take a tumour type in which 20% of

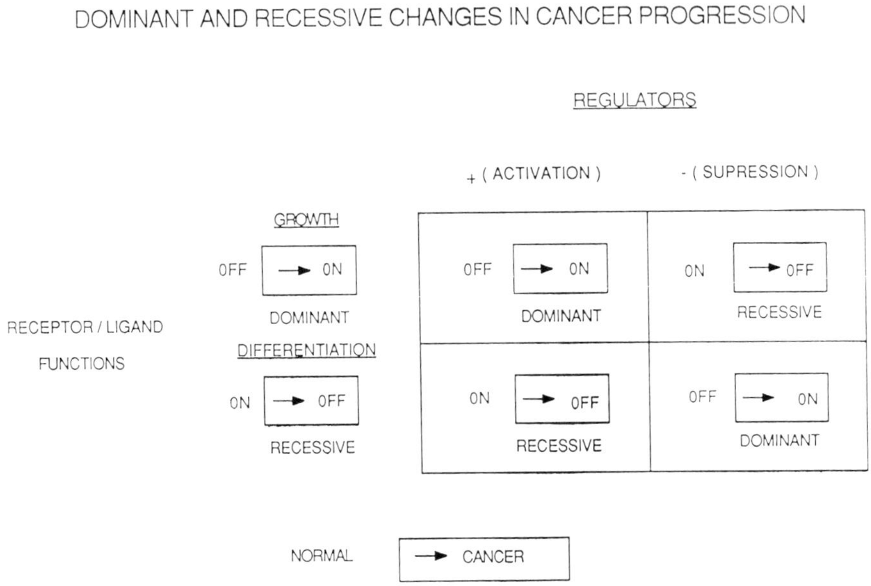


FIG. 2. (*Bodmer*) Dominant and recessive changes in cancer progression.

tumours have loss of heterozygosity on one chromosome, 20% on another chromosome, and a proportion on yet other chromosomes. Does that really mean that all those chromosomes contain genes that are likely to be involved?

*Stanbridge:* One could imagine that there are a number of redundant mechanisms, any one of which is sufficient to hold a cell in some form of homeostasis that prevents malignant growth. That to me suggests that if one did a somatic cell hybrid experiment, one would have to lose a significant amount of genetic information contributing to homeostasis before the cell becomes malignant. Yet in our experience and that of Dr Harris, those segregation events that lead to re-expression of tumorigenicity are, at the chromosomal level, discrete and rather specific. Different chromosomes seem to be involved in the control of malignant expression of different cancers, but in each case one given chromosome always seems to be involved. In our human hybrid cell system the chromosome constitution of the hybrids is very stable—there is a lot of normal genetic information still in the cell, which is once again tumorigenic.

*Sachs:* The experiments on the reversion of malignancy by chromosome changes in malignant cells without cell fusion that I mentioned also showed that malignant cells still have normal genetic information that can suppress malignancy.

*Harris:* In naturally occurring tumours, it is, of course, very difficult to say whether a particular chromosomal aberration has some meaning unless you find a constant association between a particular aberration and a particular tumour. In experimental systems, it isn't so difficult. Chromosomal changes are occurring all the time in cells in culture, but if they confer a selective advantage *in vivo*, they go to fixation after a couple of passages through the animal. So it is not difficult in transplantable tumours to isolate changes that do confer a selective advantage.

*Bodmer:* The quantitation is just classical population genetics of mutation, drift and selection, which also applies to asexual reproduction and so to somatic evolution in tumour progression. One needs an estimate of the population size, which is usually large, and the rate of variant production. With a low rate of variant production and a large population size most of the changes that go to fixation must be mediated by selection. In somatic evolutionary growth,a selection advantage of 10%, which may be very hard to measure in culture, could be quite significant to the outgrowth of the tumour. So genetic steps may take place whose functional significance is hard to establish *in vitro*.

*Adams:* But don't you suspect that the cell will sometimes carry along irrelevant changes that are linked to the relevant change?

*Bodmer:* You can calculate the extent to which that happens based on the frequency of variants. When you have gross aneuploidy and such a high rate of variant production, I would agree with what you are saying. There is a problem of allele loss when you are looking at grossly aneuploid karyotypes. The less that is the case, the easier it is to be sure that the particular changes are significant.

*Harris:* If you have an aneuploid cell line and measure the variability between cells, you might suppose that if you cloned the line you would reduce this variability. In fact, you increase it, perhaps because a very small number of cells having to survive in a large amount of medium are subjected to extreme and variable selective pressures. The amount of variation in the very early progeny of a clone is much greater than that seen after the clone grows up into a bulk culture.

*Wyke:* Yes, but it's conceivable that whatever mechanisms may be responsible for aneuploidy would favour the appropriate evolution and at the same time favour random events.

*Benedict:* In primary tumours there are specific chromosome changes that appear to be very important selectively as an *in vitro* phenomenon. For example, in neuroblastoma, double minutes and N-*myc* amplification are probably extremely important for the growth of that cell type in culture. It also appears to be important for the progression of neuroblastoma *in vivo*. Results obtained by Brodeur and Seeger show a strong clinical significance in the amount of N-*myc* amplification and prognosis. However, N-*myc* amplification may not be relevant to why that tumour arose.

In one neuroblastoma cell line, we tried to eliminate a large marker chromosome containing an amplified N-*myc* region, but were never successful—in several thousand metaphases examined, we were unable to find a cell which had lost the marker chromosome. I don't believe that N-*myc* was important in how that tumour arose but it is essential for its growth in cell culture.

*Klein:* We have been working with a SEWA ascites tumour, in which double minutes were originally discovered by Levan & Levan. They noted that double minutes were present as long as the tumour grew in the ascites form, but disappeared in tissue culture. Bishop's group found that this tumour has c-*myc* amplification. Recently, we repeated this and found maintenance of c-*myc* amplification as long as the tumour was passaged in the ascites form but not when the tumour was passaged in the solid form. Amplification occurred again when the tumour was re-passaged in the ascites form. So depending on the growth form, amplified c-*myc* seems to give the cell a selective advantage.

I think the maintenance or non-maintenance of episomes can also depend on the mechanics of the system. An example where maintenance of an episome is apparently of no advantage to the cell and yet is maintained is EBV. EBV-negative Burkitt's lymphoma grows well *in vitro*. Conversion with EBV leads to viral persistence in the form of multiple episomes; they persist because the virus has a special mechanism to maintain them. I don't think you could make any generalization from that.

*Klinger:* Dr Margot Kaelbling and others in our group, in collaboration with Dr Frank Lilly's group at Einstein, have been looking at MNU (N-nitroso-N-methylurea)-induced lymphomas in RF/J mice. MNU is a very potent carcinogen, producing lymphomas in 100% of injected mice within about 5–8 weeks. This makes it possible to examine the chromosomes of thymus cells before the lymphoma is evident, as well as during its progression. The main early changes are chromosome losses, suggesting that loss of a suppressor may play a role in genesis of these lymphomas. To our surprise, most of the early changes are not random, and most persist as the lymphoma develops, so that the same changes are found in tumours of different mice, but other tumours share only some or none of the same chromosome alterations, even though the lymphomas otherwise appear identical. The rearrangements found are particularly specific, they occur at the same breakpoints in different mice. Some of these are the same as those reported for other mouse neoplasias by George Klein and his associates and by others, which result in the activation of oncogenes or other putative transformation-related genes. This, the non-random nature of the changes, and the fact that the same changes occur in different tumours, make it likely that these alterations are causally related to the transformation and progression of the lymphoma. If this is true, then the message in respect to what we have been discussing, is that alterations at a restricted number of loci can result in what seems to be the same type of lymphoma, for example, analogous to changes in 5q or in 17p that have been found in association with colorectal carcinomas of different individuals. Since relatively few non-random changes were found in the lymphomas, even at early stages, we are forced to make the surprising suggestion that MNU preferentially induces transformation-associated chromosome changes. This may explain why MNU is such a strong carcinogen, but the reason for its apparent chromosome- and tissue-specificity remain obscure.

*Knudson:* Some chromosome changes are going to be bad for the cancer cell and some will be good. Felix Mittelman has talked about different courses of karyotypic evolution in tumours: convergence and divergence. If there is not a tremendous growth advantage for a particular karyotype, there will tend to be divergence. Once there is a strong selective advantage for a clone, there tends to be convergence.

*Benedict:* There are not many near diploid solid tumour cell lines. A few, for example, retinoblastomas, are close to diploid. The fibrosarcoma cell line, HT1080, is also near diploid and very stable. We subcloned these cells for over two years, trying to get karyotypic or phenotypic variants without success. Other tumours have an extremely random chromosomal pattern. However, if you find a marker chromosome, that marker chromosome is likely to be present in every cell. Then sometimes a low percentage of the cells develop additional marker chromosomes. I don't think that the new marker provides such a selective advantage that it completely overgrows the other cells, either *in vitro* or *in vivo*. One can inject, for example, $10^7$ cells of primarily non-tumorigenic cells *in vivo* and get only one tumorigenic revertant cell to grow out. But if you just randomly passage cells, you don't see one cell taking over. If you look at tumour sections from various areas, you find the cells are fairly similar karyotypically, and that random chromosome changes have occurred. The only thing I can tell from the chromosomal pattern is that they are all related to each other and were originally derived from one clonal tumorigenic cell. It is difficult enough trying to determine what the importance of the common chromosome markers is, much less to decide whether the rarer changes have any significance.

## Reference

Bodmer W 1988 Somatic cell genetics and cancer. Cancer Surv 7:2:239–250

# Index of contributors

*Non-participating co-authors are indicated by asterisks. Entries in bold type indicate papers; other entries refer to discussion contributions.*

*Indexes compiled by Liza Weinkove*

# Subject index